BIRDS – EVOLUTION, BEHAVIOR AND ECOLOGY

TRENDS IN ORNITHOLOGY RESEARCH

BIRDS – EVOLUTION, BEHAVIOR AND ECOLOGY

Additional books in this series can be found on Nova's website under the Series tab.

Additional E-books in this series can be found on Nova's website under the E-books tab.

BIRDS – EVOLUTION, BEHAVIOR AND ECOLOGY

TRENDS IN ORNITHOLOGY RESEARCH

PEDRO K. ULRICH
AND
JULIEN H. WILLETT
EDITORS

Nova Science Publishers, Inc.
New York

Copyright © 2010 by Nova Science Publishers, Inc.

All rights reserved. No part of this book may be reproduced, stored in a retrieval system or transmitted in any form or by any means: electronic, electrostatic, magnetic, tape, mechanical photocopying, recording or otherwise without the written permission of the Publisher.

For permission to use material from this book please contact us:
Telephone 631-231-7269; Fax 631-231-8175
Web Site: http://www.novapublishers.com

NOTICE TO THE READER

The Publisher has taken reasonable care in the preparation of this book, but makes no expressed or implied warranty of any kind and assumes no responsibility for any errors or omissions. No liability is assumed for incidental or consequential damages in connection with or arising out of information contained in this book. The Publisher shall not be liable for any special, consequential, or exemplary damages resulting, in whole or in part, from the readers' use of, or reliance upon, this material.

Independent verification should be sought for any data, advice or recommendations contained in this book. In addition, no responsibility is assumed by the publisher for any injury and/or damage to persons or property arising from any methods, products, instructions, ideas or otherwise contained in this publication.

This publication is designed to provide accurate and authoritative information with regard to the subject matter covered herein. It is sold with the clear understanding that the Publisher is not engaged in rendering legal or any other professional services. If legal or any other expert assistance is required, the services of a competent person should be sought. FROM A DECLARATION OF PARTICIPANTS JOINTLY ADOPTED BY A COMMITTEE OF THE AMERICAN BAR ASSOCIATION AND A COMMITTEE OF PUBLISHERS.

Additional color graphics may be available in the e-book version of this book.

LIBRARY OF CONGRESS CATALOGING-IN-PUBLICATION DATA

Trends in ornithology research / editors, Pedro K. Ulrich and Julien H. Willett.
p. cm.
Includes index.
ISBN 978-1-60876-454-9 (hardcover)
1. Ornithology. I. Ulrich, Pedro K. II. Willett, Julien H.
QL673.T74 2009
598--dc22
2009048932

Published by Nova Science Publishers, Inc. ✚ New York

CONTENTS

Preface		**vii**
Chapter 1	On the Trail of Early Birds: A Review of the Fossil Footprint Record of Avian Morphological and Behavioral Evolution *Martin G. Lockley and Jerald D. Harris*	**1**
Chapter 2	Trophic Relationships and Mechanisms of Ecological Segregation among Heron Species in the Parana River Floodplain (Birds: Ardeidae) *Adolfo H. Beltzer, Juan A. Schnack,* *Martín A. Quiroga, María de la Paz Ducommun,* *Ana Laura Ronchi Virgolini and Viviana Alessio*	**49**
Chapter 3	Reflections of Winter Season Large-Scale Climatic Phenomena and Local Weather Conditions in Abundance and Breeding Frequency of Vole-Eating Birds of Prey *Tapio Solonen*	**95**
Chapter 4	Integrating Indigenous Knowledge of Birds into Conservation Planning in New Guinea *William H. Thomas*	**121**
Chapter 5	European Bird Species Have Expanded Northwards During 1950-1993 in Response to Recent Climatic Warming *Gregorio Moreno-Rueda*	**137**
Chapter 6	The Study of Interactions between Birds and Flowers in the Neotropics: A Matter of Point of View *Márcia A. Rocca and Marlies Sazima*	**153**

Chapter 7	Radar Ornithology - the Past, Present, and Future: A Personal Viewpoint *Sidney A. Gauthreaux*	**161**
Index		**169**

PREFACE

Birds are a commonly acknowledged indicator of biodiversity. This book presents an indigenous perspective on the effects of traditional activities on birds. Moreover, birds are among the main components for plant reproduction in tropical ecosystems, hummingbirds being the most important vertebrate pollinators in the Neotropics. This book puts together different approaches and perspectives to study bird-flower interaction networks, reinforcing the idea of communities displaying high connectedness. In addition, data on the number of occupied territories and breeding frequency (active nests) of nine species of vole-eating birds of prey in Finland are examined, using generalized linear models. It was expected that the effects of global warming on various vole-eating birds of prey at high latitudes were both positive and negative, in particular due to mild winters. Thus, because temperature affects the distribution limits of many organisms, global warming may provoke an advance of distribution ranges polewards. The authors also discuss whether European birds have advanced their distribution ranges mainly northwards in response to climatic warming. Furthermore, fossil footprints provide important evidence regarding the morphology, behavior, distribution, and ecology of ancient animals. For the first time, the entire avian track record is reviewed, including its specialized ichnotaxonomy, from the Mesozoic through the Holocene. How the evidence impacts the understanding of avian evolution and ecology is discussed as well.

Chapter 1 - Fossil footprints provide important evidence regarding the morphology, behavior, distribution, and ecology of ancient animals. In recent years, the ichnological record (pertaining to fossils other than skeletal or body parts, most familiarly and commonly tracks) of major tetrapod clades has been studied intensively. The body fossil record amply demonstrates that the origin of birds lies within the theropod dinosaur lineage (making birds extant dinosaurs, in an evolutionary sense), but the ichnological record contributes much valuable information concerning behavioral shifts during both this evolutionary transition and the early diversification of birds. Here, for the first time, we review the entire avian track record, including its specialized ichnotaxonomy, from the Mesozoic (the "Age of Reptiles," 250-65 million years ago) and Cenozoic (the "Age of Mammals and Birds," 65 million years ago through the present, including the Holocene) and consider how the evidence impacts the understanding of avian evolution and ecology.

Growing evidence from both the skeletal and track records indicates that the initial avian taxonomic, morphological, and ecological radiations took place around the Jurassic-Cretaceous boundary (about 145 million years ago). Tracks similar to, and in some cases indistinguishable from, those made by modern shorebirds (Charadriiformes), small ducks (Anseriformes), small herons (Ciconiidae), and even roadrunners (Cuculiformes) appeared, and were even regionally abundant only 15-20 million years thereafter. In contrast, the oldest body fossil records of anseriforms and possibly charadriiforms occur very close to the end of the Cretaceous (roughly 70 million years ago), and later still for ciconiiforms and cuculiforms. This strongly implies that the early track makers were members of extinct, early avian lineages with which later taxa converged in foot morphology. Feeding traces associated with some of these tracks demonstrate that behaviors reminiscent of extant herons and spoonbills had also evolved by this time. However, despite significant skeletal and footprint finds, there is little correspondence between the records — few footprints match the feet of birds represented by skeletal remains. In short, the familiar morphologies and behaviors of many modern birds actually evolved convergently with many of their extinct, Mesozoic relatives. Footprints thus have the dual benefits of providing an important, and unexpected, complementary record of early avian morphological and ecological diversity while highlighting the importance of morphological and behavioral convergence.

Although the skeletal record suggests an avian taxonomic shift at the "dinosaur-killing" Cretaceous-Paleogene (K-P$_g$) boundary extinction event, the track record provides insufficient evidence to support or refute such a shift because the dominance of shorebird-like tracks continues uninterrupted from Mesozoic to Cenozoic. Early Paleogene tracks provide evidence of large, *Diatryma*- or *Gastornis*-like, ground dwelling birds in addition to typical shorebirds and waterbirds like the Eocene anseriform *Presbyornis*. Neogene tracks include those of a few large ratites and a turkey-like species; Holocene tracks include those of several species of moa. Unlike its Mesozoic counterpart, the Cenozoic avian body fossil and ichnological records correspond much more closely.

Tracks of perching birds, raptors, and other groups that do not habitually frequent wet shorelines — the most suitable environment for track preservation — are rare. Indeed, the avian track record is dominated by the footprints of shorebirds, with a minor component attributable to large flightless and cursorial forms. Nevertheless, the body of literature on fossil bird tracks is still relatively small (~200 reports), describing about 6 ichnofamilies, comprising about 38 named ichnogenera and 65 ichnospecies.

Chapter 2 - Herons are one of the best represented families in the floodplain of the Paraná River. The fact that interspecific competition constitutes the most significant factor in resources distribution is a prevailing idea in the ecological theory. According to recent studies, even though competition is important, the modeling of the community's structure results from the combined action of other factors which operate independently from the interspecific interaction. The distribution of resources is closely related to the ecological niche concept, this being the quantitative description of the organic unit requirements. It is hypothesized in our work that the studied species: *Ardea cocoi* (White-necked Heron), *Butotides striatus* (Striated Heron), *Bubulcus ibis* (Cattle Egret), *Ardea_alba* (Great Egret), *Egretta thula* (Snowy Egret), Syrigma sibilatrix (Whistling Heron), *Nycticorax nycticorax* (Black-crowned Night Heron), *Tigrisoma lineatum* (Fasciated Tiger-Heron) and *Ixobrychus involucris* (Stripe-backed Bittern), despite of the observed sympatry, have developed adaptative mechanisms of ecological segregation. This allows these species to use the resources in such a way that their diet composition (trophic sub niche) and other parameters of their ecological requirements (temporal sub niche and spatial sub niche) are differentiated. The index of relative importance was applied to calculate the contribution of each food category to the diet of each species. The trophic overlapping of the alimentary spectra, accumulated trophic diversity, alimentary efficiency, dietary selectivity, trophic spread of the niche, spatial sub niche and habitat preference were estimated. As regards trophic spectrum, even though fishes were found to be the basic diet for all four species and insects the second food category, slight differences exist between them which would establish mechanisms at the catches level. This is reinforced by the low overlapping values obtained and the lack of significance in the selectivity values. Variations concerning temporal and spatial sub niches were also obtained. Summarizing, the coexistence is mainly based on the differential utilization of the resources as basic isolation mechanisms and less subtly on space and time. Without leaving aside the usefulness of further research, we think these results provide valuable data for the understanding of the Paraná complex system dynamics.

Chapter 3 - We examined long-term (1986–2008) data on the number of occupied territories and breeding frequency (active nests) of nine species of vole-eating birds of prey in southernmost Finland, using generalized linear models. Explaining variables included wintertime and monthly large-scale climatic conditions indicated by North Atlantic Oscillation (NAO), mean winter and monthly mean ambient temperature and depth of snow cover at five local weather stations, as well as indices of autumn and spring abundance of voles at three localities within or near to the study area. The birds of prey included six site-tenacious species, of which four (*Bubo bubo, Glaucidium passerinum, Strix aluco, Strix uralensis*) were mainly sedentary and two (*Circus aeruginosus, Buteo buteo*) migratory ones, and three more or less nomadic species (*Falco tinnunculus, Asio otus, Aegolius funereus*). We expected that climatic effects were expressed in the numbers and breeding performance of birds of prey largely via their effects on highly fluctuating vole populations. In accordance with earlier findings, numbers and breeding of vole-eaters were largely governed by the abundance of small voles, confirming the suitability of our data to the present purpose. Large-scale climatic phenomena, indicating mild winter conditions, presented a nearly significant positive influence on the numbers and breeding frequency of southerly distributed site-tenacious species (*Buteo buteo, Bubo bubo, Strix aluco)*. The combined effect of vole abundance and local mean winter temperature was negative both in sedentary *Strix aluco* and nomadic *Falco tinnunculus*. High temperatures in the beginning and at the end of winter showed positive associations. Thick snow cover combined with vole abundance showed positive associations with numbers and breeding frequency of various kinds of vole-eating birds of prey. The results followed largely our expectations though the link via vole abundance was inadequately demonstrated. Our results suggest that the effects of global warming on various vole-eating birds of prey at high latitudes were both positive and negative, in particular due to mild winters. This would lead to changes in local populations and distribution ranges of species. Due to their flexible moving habits, nomadic species might be less seriously affected than site-tenacious ones that are more dependent on local resources, such as nest sites. From a local point of view and during a short period of time, however, the impact seemed to be more pronounced on nomadic species due to their sudden and drastic shifts.

Chapter 4 - It has been difficult to integrate indigenous knowledge into conservation planning. Although indigenous naturalists have accumulated generations of observations concerning their environments, stereotypes concerning their relationship to nature have frustrated attempts to involve indigenous societies in conservation. However, unencumbered by western philosophy, indigenous naturalists have been developing a dynamic view of nature that incorporates connectedness, disturbance and recovery as a normal course of events in the natural world. This non-linear view of nature has only recently emerged as scientific consensus. In this article, I argue that communication between conservationists and indigenous people can be facilitated by using indigenous knowledge of birds to identify the impacts of tradition on biodiversity. Birds are a commonly acknowledged indicator of biodiversity. Because indigenous people have a long-range perspective on the effects of human activity on avian diversity, they can provide a perspective vital to conservation planning. Drawing on ethno-ecological fieldwork with the Hewa of Papua New Guinea, this paper presents an indigenous perspective on the effects of traditional activities on birds. The Hewa describe their traditions as shaping the environment by creating a mosaic of habitats of varying diversity. I argue that the while the current lifestyle of the Hewa may not necessarily be a template for future sustainability, the Hewa view of the natural world provides insights into the potential of indigenous people to conserve their resources.

Chapter 5 - Because temperature affects the distribution limits of many organisms, global warming may provoke an advance of distribution ranges polewards. This work examines whether European birds have advanced their distribution ranges mainly northwards from 1950 to 1993 in response to climatic warming. Bird species were separated into different categories according to their distribution. The findings show that European birds advanced their distribution ranges northward more than southward. Only northernmost species showed the contrary pattern, probably because their northward expansion was constrained by the Arctic Sea. Birds from the central Europe advanced their distribution ranges primarily northward, strongly suggesting that the change in distribution was occasioned by climate warming, as a change due to other causes predicts equal frequency of species advancing southwards and northwards.

Chapter 6 - Birds are among the main components for plant reproduction in tropical ecosystems, with hummingbirds being the most important vertebrate pollinators in the Neotropics. Flower-visiting birds of another groups (the perching birds) are often considered as parasites of the flower-hummingbird relationships. These birds do not present a high degree of specialization for nectarivory, although nectar should be a very important component of the diet of some groups. Birds usually also visit non-ornithophilous flowers, as they look for resources in flowers adapted to pollination by other animals as well. However, very few studies have focused on non-ornithophilous flowers, which means looking at the whole group of species visited by birds, from the bird's point of view— the *resource approach*. While visiting non-ornithophilous flowers, birds (usually hummingbirds) may act merely like robbers, thieves or even co-pollinators. Therefore, when the aim of the study is pollination, one should not only look for ornithophilous flowers, but also for other possible bird pollinated species, from the flower's point of view—the *pollination approach*. Studies focusing on this last approach are even scarcer at the community level. It is important to realize that the set of *ornithophilous species* are inside the wider set of *pollinated species* by birds, and this one is contained inside the set of *visited species* by birds. Studies that only pick up ornithophilous species from a community are not focussing on pollinated species by birds, but rather on a subset of that. Another problem of point of view is that most studies in the Neotropics are ground based, which may reduce sampling of canopy species information. Observation positions within the canopy greatly enhance this kind of study and should be used more often. As flowers pollinated by perching birds may be more common in the canopies of Neotropical forests, perching bird flowers and their visitors and pollinators are underestimated in communities sampled only from the ground, which means that the majority of the studies on bird-flower interactions in Neotropical forests present good lists of bird and plant species, but very incomplete interaction networks. After putting together these different approaches to study the bird-flower interaction network, we could maybe build—with the help of other animal-flower networks—a picture of a combined model of nested compartments to the whole community, connecting all animal-flower networks by interactions of co-pollination or just visits, reinforcing the idea of communities displaying high connectance.

Chapter 7 – The Beginning of Radar Ornithology: Eric Eastwood (March 12, 1910–October 6, 1981) was one of the first to use radar to study the movement of birds, and many of his studies and those of other pioneers are summarized in his book, *Radar Ornithology* published in 1967. He was elected as Fellow of the Royal Society 1968. In the *Biographical Memoirs of Fellows of the Royal Society,* Vol. 29 (November 1983), p. 177–195, F.E. Jones said the following about Eric Eastwood: "An observation that was to prove of great interest to Eastwood in later years was made by operators at a very early CHL type radar station installed at Happisburgh, on the Norfolk coast. Some echoes were positively identified as coming from a flock of geese crossing the sea. This observation, made in 1940, was the first record of the flight of birds being followed by radar and it led to extensive investigations by Eastwood in later years and to the publication of a book on the subject (Eastwood, 1967)." The CHL (Chain Home Low) radar system was developed to counter the low-level air defense threat to the United Kingdom in 1939.

In: Trends in Ornithology Research
Editors: P. K. Ulrich and J. H. Willett, pp. 1-47

ISBN: 978-1-60876-454-9
© 2010 Nova Science Publishers, Inc.

Chapter 1

ON THE TRAIL OF EARLY BIRDS: A REVIEW OF THE FOSSIL FOOTPRINT RECORD OF AVIAN MORPHOLOGICAL AND BEHAVIORAL EVOLUTION

*Martin G. Lockley[1]*and Jerald D. Harris[2]***

[1]Dinosaur Tracks Museum, University of Colorado at Denver, Denver, CO 80217, United States
[2]Dixie State College, 225 South 700 East, St. George, UT 84770 United States

ABSTRACT

Fossil footprints provide important evidence regarding the morphology, behavior, distribution, and ecology of ancient animals. In recent years, the ichnological record (pertaining to fossils other than skeletal or body parts, most familiarly and commonly tracks) of major tetrapod clades has been studied intensively. The body fossil record amply demonstrates that the origin of birds lies within the theropod dinosaur lineage (making birds extant dinosaurs, in an evolutionary sense), but the ichnological record contributes much valuable information concerning behavioral shifts during both this evolutionary transition and the early diversification of birds. Here, for the first time, we review the entire avian track record, including its specialized ichnotaxonomy, from the Mesozoic (the "Age of Reptiles," 250-65 million years ago) and Cenozoic (the "Age of Mammals and Birds," 65 million years ago through the present, including the Holocene) and consider how the evidence impacts the understanding of avian evolution and ecology.

Growing evidence from both the skeletal and track records indicates that the initial avian taxonomic, morphological, and ecological radiations took place around the Jurassic-Cretaceous boundary (about 145 million years ago). Tracks similar to, and in some cases indistinguishable from, those made by modern shorebirds (Charadriiformes), small ducks (Anseriformes), small herons (Ciconiidae), and even roadrunners (Cuculiformes) appeared, and were even regionally abundant only 15-20 million years thereafter. In contrast, the oldest body fossil records of anseriforms and possibly

---

* Corresponding author: E-mail: Martin.Lockley@cudenver.edu
** E-mail: jharris@dixie.edu

charadriiforms occur very close to the end of the Cretaceous (roughly 70 million years ago), and later still for ciconiiforms and cuculiforms. This strongly implies that the early track makers were members of extinct, early avian lineages with which later taxa converged in foot morphology. Feeding traces associated with some of these tracks demonstrate that behaviors reminiscent of extant herons and spoonbills had also evolved by this time. However, despite significant skeletal and footprint finds, there is little correspondence between the records — few footprints match the feet of birds represented by skeletal remains. In short, the familiar morphologies and behaviors of many modern birds actually evolved convergently with many of their extinct, Mesozoic relatives. Footprints thus have the dual benefits of providing an important, and unexpected, complementary record of early avian morphological and ecological diversity while highlighting the importance of morphological and behavioral convergence.

Although the skeletal record suggests an avian taxonomic shift at the "dinosaur-killing" Cretaceous-Paleogene (K-P$_g$) boundary extinction event, the track record provides insufficient evidence to support or refute such a shift because the dominance of shorebird-like tracks continues uninterrupted from Mesozoic to Cenozoic. Early Paleogene tracks provide evidence of large, *Diatryma*- or *Gastornis*-like, ground dwelling birds in addition to typical shorebirds and waterbirds like the Eocene anseriform *Presbyornis*. Neogene tracks include those of a few large ratites and a turkey-like species; Holocene tracks include those of several species of moa. Unlike its Mesozoic counterpart, the Cenozoic avian body fossil and ichnological records correspond much more closely.

Tracks of perching birds, raptors, and other groups that do not habitually frequent wet shorelines — the most suitable environment for track preservation — are rare. Indeed, the avian track record is dominated by the footprints of shorebirds, with a minor component attributable to large flightless and cursorial forms. Nevertheless, the body of literature on fossil bird tracks is still relatively small (~200 reports), describing about 6 ichnofamilies, comprising about 38 named ichnogenera and 65 ichnospecies.

INTRODUCTION

The study of the fossil footprints of birds, known technically as "avian paleoichnology" (from the Latin *palaeos*, meaning "old," and *ichnos*, meaning "a trace") might, at first glance, seem like an obscure and highly specialized sub-discipline in paleontology. Moreover, the field may be wholly unfamiliar to modern ornithologists. However, fossil footprints are surprisingly abundant, and the discipline of ichnology, practiced by ichnologists (or, more colloquially, "trackers"), has grown consistently throughout the development of modern paleontology. Following unpublished reports of purported fossil bird tracks in New England in 1802, the subject of paleornithology has been of considerable interest to mainstream paleontology and studies of macroevolution. Consequently, both the fossilized skeletal remains and footprints of birds and bird-like taxa (notably dinosaurs, or what are now commonly referred to as non-avian dinosaurs, meaning all dinosaurs other than birds) have attracted the attention of many of the most prominent paleontologists in history. For example, British zoologist Richard Owen (1804-1892) famously described the bones of the giant New Zealand moa *Dinornis* (meaning "terrible bird"; Owen, 1842a) in the same year he coined the term Dinosauria (meaning "terrible lizard"; Owen, 1842b), not long before moa footprints were described (Gillies, 1872; Williams, 1872) and explicitly attributed to *Dinornis* (Owen, 1879; see Lockley et al. [2007a] for summary). Indeed, some historians hold that moa studies "established independent New Zealand science" when local naturalists were able to "wrest control of moa research away from Richard Owen and his British colleagues" (Thode, 2008, p. 1-2). Shortly after Owen's recognition of the moa, French zoologist Étienne Geoffroy Saint Hilaire (1851) described the 3 m-tall elephant bird *Aepyornis maximus* from Madagascar. These giant birds had purportedly been seen as recently as the 17th and 18th centuries, but no scientific record of footprints made by these animals as extant track makers is known.

Following the 1861 discovery of *Archaeopteryx*, the oldest known bird, in the Late Jurassic-age (approximately 145 million years ago) Solnhofen Limestone of Germany, the debate about bird origins was catapulted into the limelight amid the already considerable excitement surrounding the publication of *The Origin of Species* (Darwin, 1859). In opposition to Owen's anti-evolutionary stance, Thomas Henry Huxley, Darwin's famous "Bulldog," was the first to widely outline evidence linking birds and reptiles, especially some dinosaurs (Huxley, 1867, 1868, 1870; see Switek [in press] for discussion), a position much debated thereafter but supported today by a massive body of evidence and nearly universally accepted by evolutionary scientists.

Between 1836 and 1861, New England geologist and reverend Edward Hitchcock was establishing a well-earned reputation as the father of fossil footprint studies, now usually referred to as vertebrate ichnology or vertebrate paleoichnology. He published a series of classic papers and monographs on the Early Jurassic (approximately 200 million year old) fossil footprints of the Connecticut Valley region (e.g., Hitchcock, 1858, 1861) and amassed a huge collection that is still housed today at Amherst College, Massachusetts. In his first paper, Hitchcock (1836) coined the blatantly ornithological name *Ornithichnites* for these "stony bird tracks" that were, in light of the widespread Biblical literalism of the time, often referred to as the tracks of "Noah's raven" despite the fact that many were far too large and robust to have been made by any corvid. The erection of this formal name initiated the ichnological practice of applying Linnean binomials to trace fossils — *Ornithichnites* is what trackers refer to as an "ichnogenus." Throughout his career, Hitchcock inferred that almost all the footprints from the Connecticut Valley were those of ancient birds. Even though large New Zealand and Madagascan bird fossils were not of Jurassic age, such a conclusion seemed consistent with the discovery of the moa and elephant bird, and many of Hitchcock's most eminent contemporaries, including Darwin himself, wrote in support of his conclusions. Hitchcock regarded the aforementioned discovery of the Jurassic *Archaeopteryx*, just three years before his death in 1864, as vindication of his interpretations. He later subdivided the many track morphologies encompassed under the "*Ornithichnites*" label with different names; the most common track type was called *Grallator* (Figure 1) based on the assumption that they were made by waders of the avian Order Grallae (an antiquated, artificial grouping containing long-legged birds, such as sandpipers and herons, as well as flightless, cursorial birds, such as ostriches). The largest three toed tracks, which he named *Eubrontes* (meaning "true thunder"), were larger than any that could be attributable to living birds, but quite in line with moa or elephant bird proportions (Figure 2). It was not until after Hitchcock's death — indeed, not until after the recognition of the Dinosauria as an entity — that the Connecticut Valley tracks were attributed to dinosaurs. In fact, the first published attribution of a fossil track to a dinosaur did not come until 1862, when giant, three-toed tracks from England were first tentatively attributed to the dinosaur *Iguanodon* rather than to giant birds (Delair, 1989).

Figure 1. *Grallator*, a common Early Jurassic dinosaur track originally named by Hitchcock (1858) to denote a similarity to birds of the paraphyletic group Grallae.

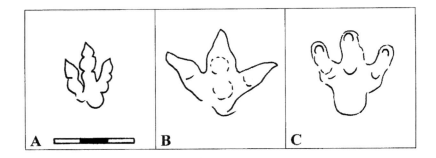

Figure 2. Dinosaur, *Dinornis*, and *Diatryma*: a gallery of "terrible" tracks made by giant birds and bird-like forms. **A**: The dinosaurian track *Eubrontes* from the Early Jurassic; **B**: track of the large moa *Dinornis* from the Holocene of New Zealand; **C**: track of a large, ground dwelling bird, possibly *Gastornis* (= *Diatryma*).

Thus, the 19th century history of research on bird tracks is inextricably linked with both the discoveries of skeletal remains of giant birds that had only recently become extinct and much older dinosaur tracks with distinctly bird-like characteristics. Put another way, the disciplines of paleornithology and avian paleoichnology have equally long, entwined histories that place them squarely at the center of important evolutionary debates concerning bird origins. At present, the link between paleornithology and avian paleoichnology remains strong because the two avenues of research complement each other via an ever-growing database on the evolution and distribution of fossil birds based on both skeletal remains and footprints. Interestingly, each discipline sometimes provides evidence that suggests somewhat different interpretations of avian evolution. This database gives us a much improved understanding of avian evolution and early diversification and demonstrates the important, if niche, role played by the study of footprints.

NAMING AND CLASSIFYING FOSSIL BIRD TRACKS

The classification of trace fossils (ichnotaxonomy) is a specialized discipline. Stated technically, ichnology is necessarily parataxonomic, meaning that ichnites are named and classified (as ichnotaxa) entirely separately from the organisms (taxa) that produce them, a convention formally codified by the International Code of Zoological Nomenclature (ICZN). Ichnotaxa must be formally named in the scientific literature, based on the morphologies the traces themselves exhibit, using Linnaean binomials, type specimens, etc. in essentially the same ways as in other branches of biology and paleontology. Thus, the aforementioned ichnogenus *Grallator*, or the ichnospecies *Grallator cursorius*, is an ichnotaxonomic or parataxonomic label applied *only* to certain distinctive tracks, diagnosable by specific and consistent characteristics. It is emphatically *not* the name of the track-making animal. There are rare, historical cases that refer unequivocally to tracks and organisms by the same name, such as the tracks of *Dinornis* (Owen, 1879), but these are exceptions to the rule, and in such cases, the use of the track maker name has no formal ichnotaxonomic significance — it is merely the case that the probable track maker can be identified with a high level of confidence. As noted below, it is not necessarily the primary aim of ichnotaxonomy to identify the track maker, which may or may not be known from the skeletal (or any other) record.

Today, assigning a track morphology to an individual species is simply a matter of observing a trace maker register traces. This is the foundation of track field guides (e.g., Elbroch and Marks, 2001), and no special track names are required because the track maker of a particular track can be observed. Paleoichnology, however, lacks the ability to observe track makers, and indeed the track maker may not be known at all from skeletal remains. As a result, hypothesized affiliations between an ichnotaxon and a specific trace maker are based on inference; when appropriate, extinct or extant organisms may be selected to serve as models, and *only* as models, to understand track characteristics and trackway dynamics. Fitting of the track to the track maker has been dubbed the "Cinderella Syndrome" (Lockley, 1998), but as detailed below, this is not the primary aim of ichnology — the science produces much other useful data independent from the ability to identify a track maker. Only in very rare circumstances are deceased track maker body fossils preserved at the ends of trackways (most famously horseshoe crabs from the Upper Jurassic Solnhofen Limestone of Germany [Barthel et al., 1990, fig. 5.5]). In such instances, an ichnotaxon can be attributed to a specific taxon with certainty, but even then the name of the taxon and the trace must be kept separate because there may be other taxa capable of registering identical traces. Biogeographic isolation of taxa can also sometimes narrow trace maker candidates of specific trace types to a single taxon, as in the case of the New Zealand moa (Lockley et al., 2007a) or the indigenous Mallorcan goat (Fornós et al., 2002).

Ichnotaxon names frequently are constructed to imply a particular track maker (or group of track makers), a practice intended to use the morphology of a known track maker as a model for the type of appendage that made the track, but this practice does not indicate definite, or even probable, affinities. For example, the name of the dinosaur track *Tyrannosauripus* suggests, superficially, that it was made by the theropod dinosaur *Tyrannosaurus*, but this is not based on direct association between any known track specimen and a skeleton of *Tyrannosaurus*. Instead, it is possible that the track maker was not

Tyrannosaurus but another large theropod dinosaur that had similar feet — *Tyrannosauripus* does not equal *Tyrannosaurus*, but it does equal a morphology resembling that of a *Tyrannosaurus* foot. It is important to keep this distinction in mind when considering avian footprints because many have names similar to those of particular avian taxa, but they do not unequivocally mean that the tracks were indeed made by members of those groups.

Table 1. Purported Cretaceous avian ichnotaxa listed in chronological order of naming, including 19 ichnospecies accommodated in 15 ichnogenera (in bold). Ichnospecies marked with an asterisk (*) are here interpreted as pertaining to non-avian dinosaurs. Indented, non-bold ichnospecies are assigned to previously named ichnogenera. For a complete list of Mesozoic bird tracks, *Gruipeda dominguensis* (de Valais and Melchor, 2008), of uncertain age, can be added to the list below.

Cretaceous avian ichnotaxa	Locality	References
Ignotornis *mcconnelli*	Colorado, USA	Mehl, 1931
Koreanaornis *hamanensis*	Korea	Kim, 1969
Aquatilavipes *swiboldae*	Canada	Currie, 1981
Yacoriteichnus *avis*	Argentina	Alonso and Marquillas, 1986
Jindongornipes *kimi*	Korea	Lockley et al., 1992
Patagonichornis *venetiorum*	Argentina	Leonardi, 1994
Aquatilavipes *sinensis*	Sichuan, China	Zhen et al., 1995
Hwangsanipes *choughi*	Korea	Yang et al., 1995
Uhangrichnus *chuni*	Korea	Yang et al., 1995
Archaeornithipus *meijidei**	Spain	Fuentes Vidarte et al., 1996
Magnoavipes *lowei**	Texas, USA	Lee, 1997
Aquatilavipes curriei	Alberta, Canada	McCrea and Sarjeant, 2001
Aquatilavipes izumiensis	Japan	Azuma et al., 2002
Barrosopus *slobodai*	Argentina	Coria et al., 2002
Sarjeantopodus *semipalmatus*	Wyoming, USA	Lockley et al., 2004
Shandongornipes *muxiai*	Shandong, China	Li et al., 2005
Ignotornis *yangi*	Korea	Kim et al., 2006
Goseongornipes *markjonesi*	Korea	Lockley et al., 2006a
Pullornipes *aureus*	Liaoning, China	Lockley et al., 2006b

With invertebrates, trace fossil morphologies may be largely independent of the morphological features of the potential trace makers (Bromley, 1996; Lockley 2007). However, with well-preserved vertebrate footprints, basic foot morphology (e.g., size, number, dimensions, orientation, and arrangement of toes and pad impressions, step and stride characteristics, and, in well-preserved examples, even skin traces) can be easily identified. In describing tracks, ichnologists typically employ terms like "didactyl," "tridactyl," and "tetradactyl" in much the same way as would an ornithologist describing actual feet. Likewise, terms such as "anisodactyl," "pamprodactyl," and "zygodactyl" can be applied equally to the footprint or the foot itself.

On the Trail of Early Birds: A Review of the Fossil Footprint Record... 7

Table 2. Purported Cenozoic avian ichnotaxa listed in chronological order of naming, including 45 ichnospecies accommodated in 23 ichnogenera.

Cenozoic avian ichnotaxa	Locality	References
Ornithichnites argenterae	Italy	Portis, 1879
Urmiornis *abeli*	Iran	Lambrecht, 1938,
= *Iranipeda abeli**	Iran	emended Vialov, 1989
= *Gruipeda abeli*	Iran	emended Sarjeant and Langston, 1994
Ardeipeda *egretta*	Romania	Panin and Avram, 1962
Ardeipeda gigantea	Romania	Panin and Avram, 1962
Ardeipeda incerta	Romania	Panin and Avram, 1962
Gruipeda *maxima*	Romania	Panin and Avram, 1962
Charadriipeda *recurvirostrioidea*	Romania	Panin and Avram, 1962
Charadriipeda minima	Romania	Panin and Avram, 1962
= *Gruipeda minima*	Romania	emended Sarjeant and Langston, 1994
Charadriipeda disjuncta	Romania	Panin and Avram, 1962
= *Gruipeda disjuncta*	Romania	emended Sarjeant and Langston, 1994
Charadriipeda becassi	Romania	Panin and Avram, 1962
= *Gruipeda becassi*	Romania	emended Sarjeant and Langston, 1994
Anatipeda *anas*	Romania	Panin and Avram, 1962
Gruipeda minor	Romania	Panin, 1965
Gruipeda intermeduia	Romania	Panin, 1965
Avipeda *phoenix*	Ukraine	Vialov, 1965
Avipeda sirin	Ukraine	Vialov, 1965
Avipeda filiportalis	Ukraine	Vialov, 1965
= *Gruipeda filiportalis***	Ukraine	emended Sarjeant and Langston, 1994
Antarctichnus *fuenzalidae*	Antarctica	Covacevich and Lamperein, 1970
Ludicharadripodiscus *edax*	France	Ellenberger, 1980
Reyesichnus *punensis*	Argentina	Alonso et al., 1980
Aviadactyla *media*	Hungary	Kordos, 1985
Ornithotarnocia *lambrecti*	Hungary	Kordos, 1985
Tetraornithopedia *tasnadii*	Hungary	Kordos, 1985
Passeripedia *ipolyensis*	Hungary	Kordos, 1985
Phenicopterichnum *rector*	Argentina	Aramayo and Manera de Bianco, 1987
Carpathipeda *panini*	Hungary	Kordos and Prakfalvi, 1990
Carpathipeda vialovi	Hungary	Kordos and Prakfalvi, 1990
Gruipeda calcarifera	Texas, USA	Sarjeant and Langston, 1994
*Gruipeda becassi**	Romania	Sarjeant and Langston, 1994
*Gruipeda disjuncta**	Romania	Sarjeant and Langston, 1994
*Avipeda ipolyensis**	Hungary	Sarjeant and Langston, 1994
Avipeda adunca	Texas, USA	Sarjeant and Langston, 1994
Fuscinapeda *sirin**	Texas, USA	Sarjeant and Langston, 1994
Fuscinapeda meunieri	Texas, USA	Sarjeant and Langston, 1994
Fuscinapeda texana	Texas, USA	Sarjeant and Langston, 1994
Presbyorniformipes *feduccii*	Utah, USA	Yang et al., 1995
Gruipeda diabloensis	California, USA	Remika, 1999
Leptoptilostipus *pyrenaicus*	Spain	Payros et al., 2000
Roepichnus *grahami*	Spain	Doyle et al., 2000
Avipeda thrinax	California, USA	Sarjeant and Reynolds, 2001
Avipeda gryponyx	California, USA	Sarjeant and Reynolds, 2001
*Avidactyla panini**	Hungary	Sarjeant and Reynolds, 2001
*Avidactyla vialovi**	Hungary	Sarjeant and Reynolds, 2001
Alaripeda *lofgreni*	California, USA	Sarjeant and Reynolds, 2001
Anatipeda californica	California, USA	Sarjeant and Reynolds, 2001
Anatipeda alfi	California, USA	Sarjeant and Reynolds, 2001
Culcitapeda *ascia*	California, USA	Sarjeant and Reynolds, 2001
Culcitapeda tridens	California, USA	Sarjeant and Reynolds, 2001
Culcitapeda eccentrica	California, USA	Sarjeant and Reynolds, 2001
Ornithoformipes *controversus*	Washington, USA	Patterson and Lockley, 2004
Gruipeda lambrechti	Iran	Mirzaie Ataabadi and Khazee, 2006
Pavoformipes *pintasotoi*	Mexico	Lockley and Delgado, 2007

Bold print indicates newly named ichnogenera; an asterisk (*) indicates ichnospecies transferred to other ichnogenera; double asterisks (**) denotes that *Gruipeda filiportalis* (sensu Sarjeant and Langston, 1994) may be better accommodated in *Ardeipeda filiportalis* (see Figure 8).

With this introduction to vertebrate paleoichnological conventions, we may now proceed to outline some of the more important discoveries. In the sections that follow, we generally only employ the ichnogenus name, except where two or more distinctive ichnospecies are accommodated within a single ichnogenus. Lists of all formally named Mesozoic (Cretaceous) and Cenozoic avian ichnotaxa are given in Tables 1 and 2, respectively. It is worth noting here that the duration of the Mesozoic is far longer than the Cenozoic. Even though birds did not evolve until the late Mesozoic, about 150 million years ago (Figure 3), the duration of the Cretaceous Period (80 million years) is still longer than the entire Cenozoic Era (65 million years) (Figure 4) — more than half of avian evolutionary history is therefore encompassed in the Mesozoic.

A REVIEW OF IMPORTANT FOSSIL BIRD FOOTPRINT DISCOVERIES

Early perceptions and reports of fossil bird tracks: 1830s-1920s

In reviewing the track record of birds, two conventional approaches, both appropriate to the inherently historical nature of paleontology, can be considered. The first, followed in this section, is to review the history of discovery, and how it progressively shaped interpretations in the field. The second, further explored below, is to review all that is currently known in order to construct an evolutionary synthesis keyed to the geologic timescale.

Discounting early claims that tracks, such as Hitchcock's Early Jurassic *Grallator* and *Eubrontes*, were made by true birds rather than non-avian dinosaurs, the first report of unequivocal fossil bird tracks was made by Desnoyers (1859), who described large tracks from Upper Eocene (~37-34 million year old), gypsum-bearing lake deposits of the Paris basin, France. Desnoyers was struck by the large size (length ~40 cm) of these tracks and compared them with the Jurassic forms described by Hitchcock from the Connecticut Valley. Although the French tracks were never illustrated, Buffetaut (2004) agreed with Desnoyers that they may have been made by the giant, ground dwelling bird *Gastornis* (a senior synonym of the more famous *Diatryma* reported from somewhat older Eocene deposits in North America by the famous paleontologist Edward Drinker Cope [1876]). More than 50 years passed before Meunier (1906) illustrated quite different, and much smaller, tracks of shorebirds from the same gypsum beds (Figure 2), and almost another century before these were named in honor of Menuier as *Fuscinapeda meunieri* by Sarjeant and Langston (1994). This earliest report (Desnoyers, 1859) of what appear to have been the first true bird tracks ever found constituted what Buffetaut (2004) called an unsolved "riddle" because the youngest gastornithid bird fossils were older than the Paris basin tracks. This enigma was only partly resolved by the finding and naming of another, slightly older *Gastornis*-like track from Eocene deposits of Washington state (Patterson and Lockley, 2004).

The next chapter in the history of avian paleoichnology began with the aforementioned discovery of moa tracks in New Zealand in the 1860s. Many of these tracks, first described in the late 1800s (Gillies, 1872; Williams, 1872; Cockburn-Hood, 1874; Owen, 1879; Voy,

1880; Hill, 1895), were collected from mostly soft, friable, volcanic ash-rich sediment. Along with numerous additional footprints excavated in 1912 (Hill, 1913; Lockley et al., 2007a), many were distributed to major museums in New Zealand and North America. As a result, a representative sample of moa tracks has been preserved for scientific study.

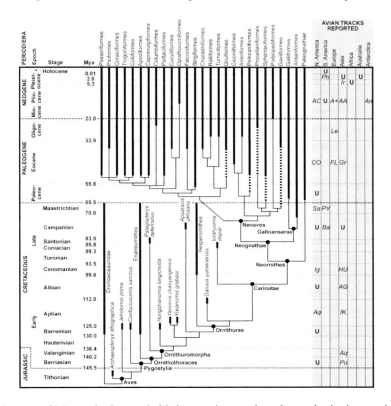

Figure 3. Cladogram of Mesozoic-Cenozoic birds superimposed on the geologic time scale, showing both taxon ranges, based on skeletal remains, and geographically and chronologically similar bird footprint occurrences (shaded columns). Thick line segments = known range occurrences; dashed line segments = possible range extensions based on specimens of debated phylogenetic affinities; thin line segments = ghost lineage extensions inferred from the phylogeny and earliest occurrences of less inclusive taxa. (Mesozoic portion of diagram modified from You et al. [2006]; phylogeny of Neornithes from Livezy & Zusi [2007a, b] with emendations and range data from Mayr [2009]). *A+* = *Anatipeda, Ardeipeda, Aviadactyla, Carpathipeda, Charadriipeda, Gruipeda, Ornithotarnocia, Passeripedia, Roepichnus,* and *Tetraornithopedia; AA* = *Antarctichnus, Avipeda,* and *Culcitapeda; AC* = *Alaripeda, Culcitapeda, Pavoformipes,* and *Reyesichnus; AG* = *Aquatilavipes, Goseongornipes, Jindongornipes, Koreanaornis, Shandongornipes,* and *Uhangrichnus; Aq* = *Aquatilavipes; Ba* = *Barrosopus; CO* = *Ardeipeda, Charadriipeda, Fuscinapeda, Ornithoformipes,* and *Presbyornithiformipes; Gr* = *Gruipeda; FL* = *Fuscinapeda* and *Ludicharadripodiscus; HU* = *Hwangsanipes* and *Uhangrichnus; Ig* = *Ignotornis; IK* = *Ignotornis* and *Koreanaornis; Ir* = *Iranipeda; Le* = *Leptoptilostipus; Ph* = *Phoenicopterichnum; Pu* = *Pullornipes; PY* = *Patagonichornis* and *Yacoriteichnus; Sa* = *Sarjeantopodus;* U = unnamed or unattributed (also known from many localities with and without named or attributed tracks).

The naming of the first bird tracks: 1930s

Despite the excitement surrounding the discovery of bird-like dinosaur tracks, possible *Gastornis* tracks, and moa skeletal remains and footprints, no formal names were assigned to any bird tracks between the 1830s and the early 20th century. Although Portis (1879) applied the name *Ornithichnites argenterae* to a purported bird track from the Cenozoic of Italy, it is not clear that the specimen he described was an actual track; Sarjeant and Reynolds (2001) referred to it as " a sedimentary structure of non-biological origin." It was not until the 1930s that the first bird track was formally named. This honor fell to a sample of well-preserved bird tracks, named *Ignotornis* by Mehl (1931), from 100 million year old ("mid" Cretaceous), coastal plain sandstones from Colorado that also yield abundant dinosaur tracks. These bird footprints (Figure 4) were not only the first formally named from the fossil record, they were also the first recognized from the Mesozoic (Table 1). *Ignotornis* is a highly distinctive track type that occurs in a large sample (~360 footprints) that has recently been restudied in detail (Lockley et al., 2009). The sample can be resolved into several dozen, well-defined, continuous trackway segments that give interesting insight into gait and behavior — as discussed below, the tracks are remarkably convergent with those of various small species of extant herons.

A few years after Mehl, Lambrecht (1938) described the second formally named bird track, *Urmiornis abeli*, from the Pliocene of Iran. These ~2-5 million year old tracks have since been renamed (because *Urmiornis* refers to a gruiform taxon based on body fossils) as *Iranipeda abeli* (Vialov, 1989). Excluding the dubious application of Hitchcock's appellation *Ornithichnites* to an Eocene track (Portis, 1879; Sarjeant and Reynolds, 2001), the Iranian track was the first formally named Cenozoic bird footprint (Table 2). Following these initial forays into avian ichnotaxonomy, there was lull in activity until the 1960s.

The 1960s: attempts to standardize Cenozoic avian ichnotaxonomy

Panin and Avram (1962) studied Miocene footprints from the Carpathian region of Romania; their resulting publication is a significant landmark in avian paleoichnology: see Brustur (1997) for review. Although bird tracks had been reported from the region by Grozescu (1918), they had not been named. Panin and Avram (1962) erected, described, and illustrated the following new ichnogenera (Figure 5) with inferred track makers (in parentheses): *Ardeipeda* (Ardeidae), *Gruipeda* (Gruidae), *Charadriipeda* (Charadriidae), and *Anatipeda* (Anatidae). Note that all the ichnogenera are named after presumed track makers, identified at the "family level." As noted above, such naming conventions are not ideal, but this was not well understood at the time of their work. Moreover, they are somewhat more understandable in instances such as the Carpathian tracks because relatively modern avifaunas were in place during the Miocene (Figure 3). The ichnogenus names also have the consistent suffix "-peda," from the Latin *pedis*, meaning foot; this suffix (and variations thereof) is frequently used for vertebrate ichnotaxa. This convention became popular with eastern European researchers; it was also adopted in descriptions of mammal tracks by Vialov (1965) (e.g., *Bestiopeda* for carnivoran tracks and *Rhinoceripeda* for rhinocerotid tracks, also from the Carpathian region of the Ukraine). Panin (1965) named additional *Gruipeda* ichnospecies. Vialov (1965) also introduced the inherently broad term *Avipeda*, which he ostensibly proposed to apply to all bird tracks. For reasons discussed below, as an ichnogenus *Avipeda* is problematic, not least for the obvious reason that there is wide variation in bird track morphology as already implied by the family-level distinctions already proposed and partially codified by Panin and Avram (1962).

Although the publication was not widely distributed, another significant bird track was reported by Kim (1969) from the Cretaceous of Korea and eponymously named *Koreanaornis*. This was the first bird track reported from Asia, only the second from the Mesozoic, and only the second (after *Iranipeda*) to be named in recognition of a modern nation. *Koreanaornis* was a harbinger of many subsequent and significant Cretaceous bird track discoveries in Asia.

Continuing chronologically, the shorebird track *Antarctichnus*, from the Cenozoic of King George Island, Antarctica (of the Argentinian Shetland Islands), was the only bird track named during the 1970s (Covacevich and Lamperein, 1970, 1972) and the first to be assigned a label referring to a whole continent. Significant, newly named ichnotaxa reported in the 1980s include *Ludicharadripodiscus*, a shorebird-like track from the Eocene of France (Ellenberger, 1980), *Aquatilavipes*, a shorebird-like track from the Cretaceous of Canada (Currie, 1981), and four tracks named by Kordos (1985) from the Early Miocene of Hungary: *Aviadactyla*, a shorebird-like track, *Ornithotarnocia*, a duck-like track, *Tetraornithopedia*, described as similar to *Ornithotarnocia* but with a "hind toe print" or hallux; Kordos 1985, p. 365), and *Passeripedia*, purportedly a passerine track. Also named in the 1980s were a variety of avian ichnotaxa from Argentina: *Reyesichnus* (Alonso et al., 1980), a Miocene shorebird-like track, *Yacoriteichnus*, yet another shorebird-like track, this time from the Late Cretaceous (Alonso and Marquillas, 1986), and *Phoenicopterichnum*, a flamingo-like track from the Late Pleistocene (Aramayo and Manera de Bianco, 1987).

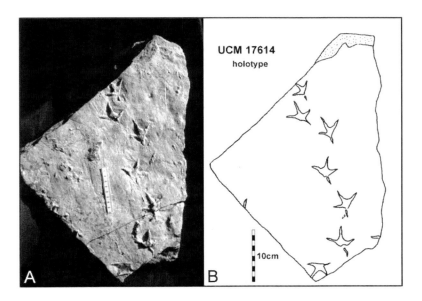

Figure 4. *Ignotornis*, the first bird track morphotype formally named and the first ever reported from the Mesozoic. It occurs in Cretaceous deposits in Colorado (after Mehl, 1931 and Lockley et al., 2009).

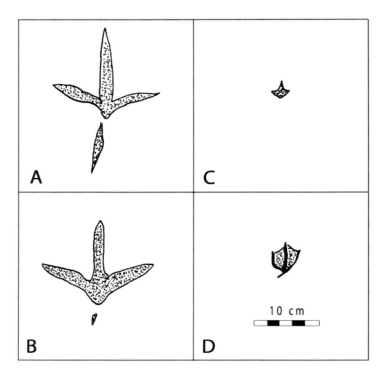

Figure 5. Four ichnogenera named after, and correlated with, extant avian groups by Panin and Avram (1962): **A**: *Ardeipeda* (Ardidae); **B**: *Gruipeda* (Gruidae); **C**: *Charadriipeda* (Charadriidae); **D**: *Anatipeda* (Anatidae). All are drawn to the same scale.

Expanding the track record of Cretaceous birds: 1990s-present

The 1990s saw a significant increase in the discovery and naming of Cretaceous bird tracks, especially from East Asia. More shorebird-like tracks were reported, including the new ichnogenus *Jindongornipes* from Korea (Lockley et al., 1992) and a purported new ichnospecies of *Aquatilavipes* from China (Zhen et al., 1995). Yang et al. (1995) named two new Cretaceous avian ichnogenera, *Hwangsanipes* and *Uhangrichnus*, from Korea; they also erected the name *Presbyornithiformipes* for a previously known Eocene trackway of a duck- or flamingo-like bird (Erickson, 1967), possibly similar to the contemporaneous *Presbyornis*. This trackway includes not only footprints, but also distinctive bill dabble traces (Figure 6). Like the much younger, and later-named, flamingo-like track *Phoenicopterichnum* (Aramayo and Manera de Bianco, 1987), *Presbyornithiformipes* was named with a very specific track maker in mind, even though it should be noted that the name translates as "*Presbyornis*-like track" rather than literally "*Presbyornis* footprint" — there may have been other birds around at the time that hypothetically also could have made the tracks.

Two track types named in the late 1990s, *Archaeornithipus* (Fuentes Vidarte et al., 1996) from the Cretaceous of Spain, and *Magnoavipes* (Lee, 1997) from the Cretaceous of Texas, represent large track makers unlike those responsible for making any of the aforementioned small bird tracks. They therefore may represent gracile, bird-like, but non-avian theropod dinosaurs such as the appropriately named ornithomimosaurs ("bird mimic" dinosaurs) (Lockley and Meyer, 2000; Lockley et al., 2001). Leonardi (1994) also illustrated tracks from the Late Cretaceous of Argentina using the label *Patagonichornis*, a name derived from an unpublished manuscript by Rodolfo Casamiquela.

The only new Cenozoic ichnogenera named in the 1990s were *Carpathipeda* (Kordos and Prakfalvi, 1990) and *Fuscinapeda* (Sarjeant and Langston, 1994). The latter was split into three ichnospecies: one to encompass one of Vialov's many Miocene ichnospecies of *Avipeda* that was deemed different enough to warrant its own ichnogenus, another to describe Late Eocene tracks from Texas, and a third, as noted above, to describe the historically significant tracks named by Meunier (1906) from the Eocene of France. Sarjeant and Langston (1994) also proposed transferring several European ichnospecies into different ichnogenera (Table 2).

A series of Cenozoic bird tracks were named between 2000 and the present. *Leptoptilostipus pyrenaicus* (Payros et al., 2000) is a distinctive, web-footed bird track (Figure 7) known since the 1960s (de Raaf, 1965) from the Oligocene of Spain. Doyle et al. (2000) erected *Roepichnus* for tracks from the Miocene of Spain, describing them as duck-like tracks similar to *Anatipeda*. Studies of Miocene tracks from California by Sarjeant and Reynolds (2001) resulted in the introduction of *Alaripeda* and *Culcitapeda*, the former based on a poorly preserved track — the result of soft substrate conditions — and the latter representing a large (11.5 cm wide) web-footed track resembling a modern goose footprint. Again in this study, Sarjeant and Reynolds (2001) proposed transferring several European ichnospecies into different ichnogenera (Table 2). Patterson and Lockley (2004) named a single, very

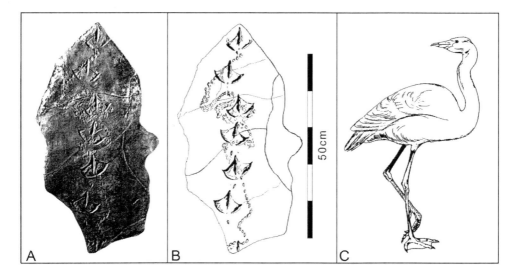

Figure 6. **A**: Photograph, and **B**: line drawing of the type specimen of the duck or flamingo-like track *Presbyorniformipes* (Yang et al., 1995). Note the distinctive bill-dabble traces. **C**: Restoration of the probable track maker, *Presbyornis* (after Erickson, 1967).

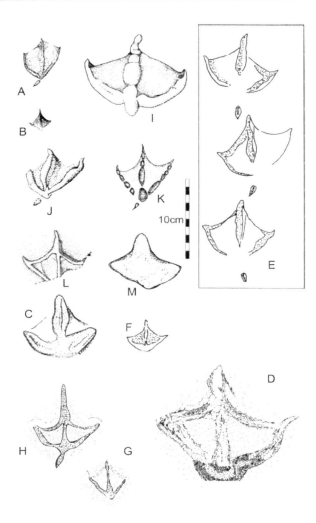

Figure 7. Tracks of web-footed track makers. **A**: *Anatapeda anas* Panin and Avram, 1962; **B**: *Charadariipeda recuvirostriodea* Panin and Avram, 1962; **C**: *Ornithotarnocia lambrechti* Kordos, 1985; **D**: *Phoenicopterichnum rector* Aramayo and Manera de Bianco, 1987; **E**: *Presbyornithiformipes feduccii* Yang et al., 1995; **F**: *Uhangriichnus chuni* Yang et al., 1995; **G**: *Roepichnus grahami* Doyle et al., 2000; **H**: *Leptoptilostipus pyrenaicus* Payros et al., 2000; **I**: *Culcitapeda eccentrica* Sarjeant and Reynolds, 2001; **J**: *Anatapeda alfi* Sarjeant and Reynolds, 2001; **K**: *Anatapeda californica* Sarjeant and Reynolds, 2001; **L**: *Culcitapeda tridens* Sarjeant and Reynolds, 2001; **M**: *Culcitapeda ascia* Sarjeant and Reynolds, 2001.

large (33 cm long) Eocene footprint from Washington State *Ornithoformipes*, and inferred that it represents a *Gastornis* (= *Diatryma*)-like bird.

Other ichnotaxa named in the last decade were new ichnospecies of the shorebird-like *Aquatilavipes* from Canada (McCrea and Sarjeant, 2001) and Japan (Azuma et al., 2002), the first named bird tracks, *Barrosopus*, from the Cretaceous of South America (Coria et al., 2002), and a semi-palmate track named *Sarjeantopodus* from the Late Cretaceous of Wyoming (Lockley et al., 2004). Especially noteworthy are discoveries in the track-rich Cretaceous sections of China and Korea. In China, Early Cretaceous strata produced the first known zygodactyl tracks, named *Shandongornipes* (Li et al., 2005), which closely resemble tracks made by the modern roadrunner *Geococcyx* (Lockley et al., 2007b). China also produced some of the oldest known bird tracks, from the earliest Cretaceous, which were placed in the new ichnogenus *Pullornipes* (Lockley et al., 2006a; Table 1). From Korea, Kim et al. (2006) named a new ichnospecies of *Ignotornis*, and Lockley et al. (2006b) erected the new ichnogenus *Goseongornipes*. The most recently named Cenozoic track is *Pavoformipes*, a turkey-like track from the Miocene of Mexico (Lockley and Delgado, 2007). Mirzaie Ataabadi and Khazee (2006) named a new species of *Gruipeda*, *G. lambrechti*, from the Eocene of Iran. Various other named and unnamed bird tracks have also been reported from the Cenozoic in the last two years (Abbassi and Shakeri, 2006; Hunt and Kelly, 2004; Krapovickas et al., 2007; Lockley et al., 2007; Lucas et al. 2007; Lucas and Schultz, 2007; Morgan and Williamson, 2007; Murelaga et al., 2007; Mustoe, 2002; Thrasher, 2007; Melchor 2009), but none of these have been assigned to new ichnotaxa.

de Valais and Melchor (2008) continued the tradition of Argentinian ichnotaxonomy by erecting the new ichnospecies *Gruipeda dominguensis* for tracks from the Upper Triassic-Lower Jurassic Santa Domingo Formation of Argentina, which the authors admitted was dated with some uncertainty. This study is noteworthy because it applied a name initially given to a Cenozoic ichnotaxon to Mesozoic specimens. As noted by Anfinson et al. (2009), this has rarely been done, except by Sarjeant and Langston (1994) and Sarjeant and Reynolds (2001). de Valais and Melchor (2008) follow Sarjeant and Langston in synonymizing, under *Gruipeda*, several of the aforementioned ichnotaxa, including *Urmiornis*, *Antarctichnus*, and some, but not all, examples of *Charadriipeda* and the Late Triassic *Trisauropodiscus* from southern Africa. However, they overlooked the suggestion of Lockley and Gierlinski (2006) that some *Trisauropodiscus* may be best accommodated in the non-avian dinosaurian ichnogenus *Anomoepus*, which is usually attributed to basal ornithischian track makers. Nevertheless, as noted below, this paper makes several other interesting observations and, as a serious attempt to standardize ichnotaxonomic procedure, it follows Sarjeant and Langston (1994) in extending the influential ichnotaxonomic nomenclature of Panin and Avram (1962) into the Mesozoic.

SYNTHESIS

The brief historical overview of avian paleoichnology just outlined may at first appear a more or less random chronicle of discovery and casual assignment of ichnotaxonomic names, especially considering the relatively restricted diversity of extant avian foot morphologies. However, although some names may be *nomina dubia* (based on material exhibiting no diagnostic features and therefore of dubious validity) or synonymous with other, previously named ichnotaxa when substrate variations and ontogeny are accounted for, most of the names are valid, and most of the more recent studies have been fairly conscientious in conducting appropriate comparative analyses. This is not to say that *all* workers have made the appropriate comparative analyses — the avian track literature suffers from the same problems encountered in the fossil footprint literature in general: i.e., many significant, mostly early, publications were in diverse languages (e.g., Russian, Hungarian, Romanian) and/or not widely distributed. Many studies have relied on comparative analyses based on illustrations and descriptions in the literature rather than detailed study of actual specimens, which in many cases may be difficult to locate, and may have even been destroyed by weathering if they were not collected or replicated.

Nevertheless, a holistic perspective reveals patterns of spatial and temporal distribution of track types that shed significant light on interesting aspects of avian evolution and paleoecology. Issues considered can be summarized in the following categories, discussed in turn below:

1. Dominant track types in the fossil record
2. Problems of ichnotaxonomic synonymy and preservation, including
   (i) Lumping, splitting, and synonymy
   (ii) The *Avipeda* problem
   (iii) Ichnofamilies and morphofamilies
   (iv) The ultimate aims of ichnotaxonomy
3. Geographical and geochronological distribution patterns
4. Behavioral interpretations
   (i) Collective paleoecology and ichnofacies perspectives
   (ii) Individual behavior
5. Consistency between the avian track and bone records
6. Evolutionary significance

Readers should keep in mind that, as a rigorous science, avian paleoichnology and ichnotaxonomy is still in its infancy. This paper is the first attempt at a comprehensive review of all known ichnotaxa (Tables 1, 2).

Dominant Track Types in the Fossil Record

Based on track morphology and understanding of ecomorphologies exhibited by extant birds, the vast majority of fossil avian footprints represent shorebird (or shorebird-like) and water-bird (or water-bird-like) track makers (Greben and Lockley, 1993). These tracks are generally associated with lake basin, fluvial floodplain, or coastal plain and marine shoreline deposits. This bias is to be expected, not just because shorebirds and waterbirds congregate in abundance in such environments, but because these environments facilitate the registration and preservation of tracks. Plovers, sandpipers, ducks, and herons are relatively common in such facies, and therefore their tracks predominate, but there is much less chance of preserving the tracks of birds whose behaviors do not include frequent terrestrial locomotion, such as perching birds and raptors. Similarly, birds that live in environments where sediment deposition — required for the burial and preservation of tracks — is infrequent (montane), or generally inaccessible (forest, open marine, etc.) are also unlikely to create tracks in favorable environments. This type of bias in the fossil record is a well-known phenomenon, affecting skeletal remains as well as footprints. The effects of size should also be noted: larger, heavier birds are more likely to impress feet into unconsolidated sediment than are small, light birds (small perching birds, for example), so even under ideal track registration conditions, many bird tracks will be neither registered nor preserved. Furthermore, ichnologists may more easily overlook small tracks. Based on such biases (discussed below), a widespread Mesozoic-Cenozoic "shorebird ichnofacies" has been defined (Lockley et al., 1994).

There is a common perception that the skeletal remains of birds are rare in the fossil record because, in comparison with large animals (e.g., many dinosaurs, ruminants), their bones are small, delicate and difficult to preserve. This is certainly true to a significant degree. The relatively recent moa bones and tracks provide a good example of this size-bias phenomenon because they are abundant relative to the skeletal remains and spoor of contemporary small species. Indeed, after the general category of shore- and water-bird morphotypes, the second most abundant track maker category is that of large, ground dwelling birds, such as ratites, including the New Zealand moa (Lockley et al., 2007a), Pleistocene, emu-like dromornithid tracks from Australia (Rich and Gill, 1976) and Tasmania (Rich and Green, 1974), and tracks of Pleistocene (Aramayo and Manera de Bianco, 2009) and Holocene (Aramayo, 2009) rhea from Argentina. *Patagonichornis* (Leonardi, 1994) might conceivably represent a tinamou-like paleognath; Cretaceous, roadrunner-like *Shandongornipes* tracks also represent a ground dwelling form.

Whereas vertebrates are hypothetically capable of registering thousands, perhaps millions, of tracks during an individual lifetime, each has but a single skeleton that may become (whole or in part) a fossil. Most bird bones (especially of small taxa), because of their thin walls (especially in highly pneumatic taxa, such as seabirds), are easily destroyed or removed during the transport processes that predominate in most burial situations (Trapani, 1998); scavengers and tissue decay also more readily destroy smaller, thinner bones than larger, heavier ones (Davis, 1997; Davis and Briggs, 1998). Thus, bias exists toward the preservation of certain skeletal elements when any elements are preserved at all. At sites where disarticulated bird bones are recovered (the most common type of site for fossil birds), certain bones are encountered more often than others. Very specific conditions are required for the preservation of delicate bird bones and complete, articulated skeletons: quiescent lacustrine and lagoonal settings in particular are known sources of such fossils. Such settings, however, are rare, so the record of such exquisite body fossils is limited to geologically brief windows from a handful of Konservat-Lagerstätten (sites of exceptional fossil preservation quality), such as the Early Cretaceous-age, lacustrine Jehol Group of northeastern China and Xiagou Formation of northwestern China, and Eocene shore deposits of the Fossil Lake Member of the Green River Formation of Wyoming and Grube Messel in Germany. Also important are a few Konzentrat-Lagerstätten (sites of exceptional quantities of fossils), such as the Eocene nearshore deposits of the London Clay of England and shoreline La Meseta Formation of Antarctica, and the inland Pleistocene La Brea tar pits of California. The abundance of tracks in some deposits provides a striking contrast to the rarity of delicate foot bones. Avian ichnotaxonomy thus enjoys one particular advantage over the body fossil record: ichnotaxonomy is nearly universal in always comparing footprints, but disarticulated bird bones from different parts of the skeleton are most often the basis of taxonomic assignment, even when common elements are not available for interspecies comparative analysis.

Problems of Ichnotaxonomic Synonymy and Preservation

Lumping, splitting and synonymy

Careful study of the avian paleoichnological literature reveals two recurring, and sometimes related, phenomena that have an impact on perceptions of ichnotaxonomic diversity. The first is the perennial problem of synonymy — the "lumping" and "splitting" of ichnotaxa. The second is the issue of preservation. Both problems require consideration in order to make interpretations about footprint distribution and diversity in space and time.

Ostensibly, the attribution of multiple ichnospecies to a given ichnogenus indicates that the ichnotaxonomist regards several footprint types as very similar, though not systemically identical. This means that the tracks exhibit a general similarity, but differ consistently across specimens in small details. A good example of this is the Cretaceous ichnogenus *Aquatilavipes* (Table 1), which is ostensibly represented by four ichnospecies. Of these, however, at least two probably belong in different ichnogenera: *A. sinensis* is dissimilar enough from the ichnogenoholotype of *Aquatilavipes*, but similar enough to *Koreanornis*, that it is likely a synonym of the latter (Lockley et al., 2008a); *A. curriei* is also unlike holotypic *Aquatilavipes* but is also unlike any currently named ichnotaxon in significant ways and therefore should be transferred to a new ichnogenus (R. McCrea, personal communication). Several Cenozoic ichnogenera also contain multiple ichnospecies: following the amendments and transfers proposed by Sarjeant and Langston (1994), Sarjeant and Reynolds (2001), and de Valais and Melchor (2008), *Gruipeda* and *Avipeda*, respectively, contain eleven and seven ichnospecies (Table 2). While it is certainly possible that each of these ichnospecies really was made by a different bird, each with slightly different pedal morphologies that manifest systemically as distinctive, measureable differences in track morphology, it is more likely that some of these ichnospecies are differentiated based either or both on extramorphological features (such as effects on track morphology created by substrate conditions, rather than differences in pedal morphology) or interpretations that will be questioned by other workers. For example, de Valais and Melchor (2008) proposed different criteria for distinguishing ichnospecies, ichnogenera, and ichnofamilies. While the aim of improved and standardized scientific communication behind these suggestions is admirable, it remains to be seen whether this will be realized in practice.

Just as ichnospecies are "lumped" into ichnogenera, so too are ichnogenera "lumped" into ichnofamilies (or morphofamilies) that imply more generalized, but nevertheless taxonomic, statements of inferred relationships. Each case must be judged on its merits; the differences between ichnospecies and ichnogenera, as well as between ichnofamilies, should ideally be clearly stated in the literature and based on careful descriptions of track morphology and comparative analysis. Unfortunately, for practical reasons, this is not always the case, especially when new ichnotaxa are named on the basis of incomplete or inaccessible material. For example, when Kim (1969) named *Koreanaornis*, he made no reference to the handful of known avian ichnogenera mostly named in the Romanian and Russian literature from the Miocene of eastern Europe (Panin and Avram, 1962; Vialov, 1965) — these works may have simply been unknown and/or unavailable to Kim. Conceivably, had he been aware of these, he may not have erected a new ichnogenus to encompass the Korean tracks. Likewise, a whole generation passed before Kim's paper was cited again as a result of new bird track discoveries in Korea (Lockley et al., 1992) — *Koreanaornis* and the ichnotaxa erected by Panin and Avram (1962) and Vialov (1965) were neither used for comparison nor cited in the intervening time period in the descriptions of other new bird track material, such as *Aquatilavipes* (Currie, 1981). Thus, as noted by Anfinson et al. (2009), the question arises as to whether tracks like *Koreanaornis* (Kim, 1969) and *Aquatilavipes* (Currie, 1981) might be synonyms of the previously named and nomenclaturally valid, but widely unknown, ichnotaxa with historical priority, such as *Avipeda* (Vialov, 1965; see Lucas [2007] for a recent review). Anfinson et al. (2009) even noted that there had never previously been any

On the Trail of Early Birds: A Review of the Fossil Footprint Record...   19

explicit attempt to compare these Mesozoic and Cenozoic tracks, though this rather fundamental omission may be based less on lack of access or exposure to bodies of literature than *a priori* assumptions that Mesozoic and Cenozoic birds differed radically in higher level taxonomy (see below) and, therefore, presumed foot morphology and paleoecology.

The Avipeda problem: a case study

It is becoming clear that some small, shorebird-like tracks from the Mesozoic and Cenozoic are very similar to each other. Two ichnospecies of *Avipeda*, *A. phoenix* and *A. sirin*, are closest in morphology to *Koreanaornis* (cf. Anfinson et al., 2009) and potentially have priority. On one hand, this inference suggests undue "splitting" at the ichnogenus level. Yet on the other hand, as previously noted, the ichnogenus *Avipeda* is problematic because it encompasses far too wide a range of avian track morphotypes — there has been too much "lumping," transforming *Avipeda* into what Sarjeant and Langston (1994, p. 12) called a "wastebasket" ichnogenus.

We therefore present a brief case study of *Avipeda* (Vialov 1965, 1966) to show how the original concept now encompasses what may justifiably be three ichnogenera, two of which are still difficult to discriminate with confidence. It is instructive to follow the "evolution" of *Avipeda* from 1965 to the present, making careful comparisons of all tracks at the same scale, based on descriptions and illustrations of original material (Figure 8). Vialov (1965, 1966) originally identified three ichnospecies of *Avipeda*: *A. phoenix*, *A. sirin*, and *A. filiportalis*. Both *A. phoenix* and *A. sirin* are small tracks (lengths of 1.6 and 2.5 cm, respectively); the latter is most comparable to *Koreanaornis* in size. *A. filiportalis* is huge by comparison (footprint length up to 19 cm) and is morphologically distinct because it possesses a caudally-oriented hallux trace — in this and other respects, it is similar to the earlier named, heron-like footprint *Ardeipeda egretta* (the type ichnospecies) and *Ardeipeda gigantea* (Panin and Avram, 1962; Lockley et al., 2007c). Thus, *A. filiportalis* is justifiably considered a synonym of *Ardeipeda*.

In their amendment of *Avipeda*, Sarjeant and Langston (1994) selected *A. phoenix* as the type ichnospecies, and suppressed the ichnogenus *Passeripeda* (Kordos, 1985) by synonymy with *Avipeda*. They also transferred *A. sirin* to a new ichnogenus, *Fuscinapeda* (*F. sirin* comb. nov.), as well as naming the "historic" specimen described by Meunier (1906) as a new ichnospecies, *F. meunieri*. They also erected a third ichnospecies of *Fuscinapeda*, *F. texana*, for footprints that appear different enough in details of size and morphology from either *F. sirin* or *F. meunieri* to warrant distinction. Two of Vialov's three *Avipeda* ichnospecies were thus transferred to other ichnogenera (*Ardeipeda* and *Fuscinapeda*).

Having made these distinctions, which initially confined the concept of *Avipeda* to the diminutive, 1.6 cm-long holotypic tracks of *A. phoenix*, Sarjeant and Langston (1994, p. 12-13) proceeded to erect three new *Avipeda* ichnospecies: *A. adunnca* (Sarjeant and Langston, 1994), with a foot length of 2.8-3.0 cm, and *A. thrinax* and *A. gryponix* (Sarjeant and Reynolds, 2001), with foot lengths of 2.3 and 1.9 cm, respectively. However, as a basis for comparison, *Koreanaornis*, which has very similar dimensions to *F. sirin*, was again overlooked in both the 1994 and 2001 papers, so the question of the relationship between *Koreanaornis* various species of *Avipeda* remains open.

Again, further study, with a broader and time-independent perspective, is called for.

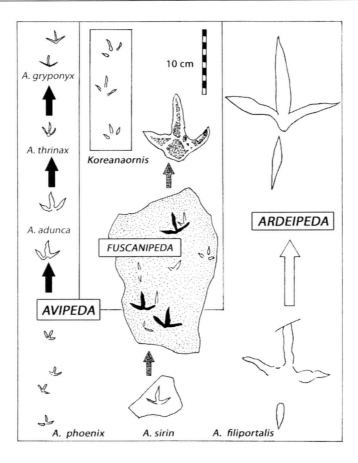

Figure 8. The evolution of *Avipeda* ichnotaxonomy 1965-2001: how one ichnogenus became three. Vialov (1965, 1966) named three ichnospecies of *Avipeda* (*A. phoenix*, *A. sirin*, and *A. filiportalis*) despite obvious differences in size and morphology. Subsequently, Sarjeant and Langston (1994) designated *A. phoenix* as the type ichnospecies for the ichnogenus and named the new ichnospecies: *A. adunca*, *A. thrinax*, and *A. gryponyx* were added later (Sarjeant and Reynolds, 2001). *A. sirin* was transferred to *Fuscinapeda* (*F. sirin*), and ichnogenus for which a second ichnospecies was erected from France (*F. meunieri*), as well as a third from Texas (*F. texana*). The utility of ichnogenus *Fuscinapeda* is questionable because *A. sirin* resembles *Koreanaornis* and it may be referable to that ichnotaxon. However, the transfer of *Avipeda filoportalis* to *Ardeipeda* is valid. All tracks drawn to the same scale. See text for details.

Ichnofamilies and morphofamilies

In addition to ichnospecies and ichnogenera, ichnofamilies (or morphofamilies) have proved a category of some utility in ichnotaxonomy. By convention, and borrowing from Linnean systematics, the suffix "-idae" has been used, as in the case of the non-avian theropod ichnofamily Grallatoridae (Lull 1904, 1953). Where an ichnogenus name ends in "-pus," the suffix "-podidae" has been used, as in Anomoepodidae and Batrachopodidae (Lull, 1904, 1953). The latter convention helps distinguish ichnofamilies from families, although some older literature, such as Lull (1904), confused the issue by simply referring to these ichnotaxa as "families." The concept of a "morphofamily" is very broad and serves only to group "forms" that are generally similar in morphology. The term "ichnofamily" has essentially the same connotation, although sometimes the suffix "-podidae" is used to more explicitly designate the role of footprints rather than broader morphotypes (cf. de Valais and Melchor, 2008).

In an apparent attempt to honor the priority of nomenclature proposed by Panin and Avram (1962) and Vialov (1965), Sarjeant and Langston (1994) proposed the "morphofamilies" Gruipedidae, Charadriipedidae, Anatapedidae, and Avipedidae. This scheme was later modified when Sarjeant and Reynolds (2001) synonymized Charadriipedidae with Anatapedidae (Table 3), a scheme modified further still by de Valais and Melchor (2008). In any event, Sarjeant and Langston (1994) overlooked the older Ignotornidae (Lockley et al., 1992), perhaps because they placed *Ignotornis* in the Gruipedidae. In our opinion, even allowing for some gradation in hallux length, there is a relatively clear distinction between the type species of *Gruipeda* (*G. maximus*), a crane-like track with a short hallux and very wide digit divarication, and *Ardeipeda* (*A. egretta*), a heron- or egret-like track with a much longer hallux and somewhat less pronounced digit divarication. Heron-like *Ignotornis*, with a long hallux (Lockley et al., 2009), is not logically allied with *Gruipeda*. In addition to overlooking Ignotornidae (Lockley et al., 1992; emended by Kim et al., 2006), Sarjeant and Langston (1994) also overlooked *Koreanaornis* (Kim, 1969) and *Jindongornipes* (Lockley et al., 1992), subsequently used by Lockley et al. (2006) as the basis of the respective, eponymous ichnofamilies Koreanaornipodidae and Jindongornipodidae. Here again the question of relationships of these ichnogenera to the various ichnofamilies proposed by Sarjeant and Langston (1994) and Sarjeant and Reynolds (2001) remains an open question that can only be resolved by further comparative analysis. While such inconsistencies complicate and, ultimately, weaken comparative analyses, it is probably fair to predict that regardless of the occasional oversight, differences of opinion are likely to arise simply as the result of new discoveries and the subjective interpretations placed on them by different workers. Given that only five Mesozoic and 11 Cenozoic ichnogenera were named prior to 1994, as compared with ~15 and ~22, respectively, by the present time, it is clear that the schemes proposed by Sarjeant and Langston (1994) and Sarjeant and Reynolds (2001) are in need of revision based on broader, more inclusive comparisons.

The problem of preservation is also much discussed in ichnology. Before naming tracks, informed decisions must be made about the quality of preservation. If tracks faithfully reflect the morphology of a track maker, they can be used to describe footprint morphology with reasonable confidence. However, if tracks were made in suboptimal substrates and/or have been compromised by erosion and weathering at any stage between their initial imprinting (registration) and subsequent exhumation, they may display various extramorphological features that are of little or no value for morphological analysis. The ichnogenus *Alaripeda* (Sarjeant and Reynolds, 2001) provides a good example. The tracks are "T"-shaped, with slight curvature to the digit traces. This configuration differs from the normal V-shape subtended by digits II and IV of the avian foot and is presumably the result of these digits (II and IV) having been bent back during progression through and over a very soft substrate, a phenomenon that has been demonstrated experimentally with emu footprints (Milàn, 2006). *Alaripeda* track morphology owes more to poor track-making and preservational conditions than to track maker foot morphology, and for this reason could be considered a *nomen dubium*. In dismissing the "morphological" validity of *Alaripeda* as a mere preservational or "extra-morphological" phenomenon, we recognize that a minority of ichnologists have argued that, in some cases, tracks that represent specific behaviors, including responses to substrate conditions, can be given formal names (Sarjeant, 1990). Indeed, de Valais and Melchor (2008) used the label *Alaripeda* to name variants in their *Gruipeda* dominated assemblage from Argentina, albeit without naming a new ichnospecies. However, as noted by Lockley (2007), the naming of extramorphological variants, even those attributed to certain distinctive behaviors, has not been a widely accepted practice. For example, different names are not proposed for tracks made by walking and running individuals. Most ichnologists are skeptical of the value of tracks named on the basis of anomalous or poorly-preserved material. We

Table 3. Grouping of ichnogenera into morphofamilies according to Sarjeant and Langston (1994), with emendations by Sarjeant and Reynolds (2001).

MORPHOFAMILY GRUIPEDIDAE	MORPHOFAMILY ANATAPEDIDAE	MORPHOFAMILY AVIPEDIDAE
Gruipeda Panin and Avram (1962)	*Anatapeda* Panin and Avram (1962)	*Aquatilavipes* (Currie, 1981)
Ardeipeda Panin and Avram (1962)	*Phoenicopterichnum* Aramayo and Manera de Bianco (1987)	*Avidactylipeda* (Kordos, 1985)
Antarchtichnus Covacevich and Lamperein, 1970		*Ornithotarnocia* (Kordos, 1985)
Ignotornis Mehl (1931)	includes **MORPHOFAMILY CHARADRIIPEDIDAE**	*Ludicharadripodiscus* (Ellenberger, 1980)
Tetraornithopoda (Kordos, 1985)	*Charadriipeda* Panin and Avram (1962)	*Fuscinapeda* (Sarjeant and Langston, 1994)
		Alaripeda Sarjeant and Reynolds (2001)

prefer the approach of Peabody (1955, p. 915), who suggested that "only the clearest records should be named specifically."

The ultimate aims of ichnotaxonomy

It is appropriate to end this section by restating that identification of a track maker is *not* the ultimate aim of ichnotaxonomy. In some cases, such as the giant moa *Dinornis*, a track-track maker correlation may be highly probable, and such "Cinderella Syndrome" solutions are satisfying (Lockley, 1998). A process of elimination should, with time, ultimately lead to improved, if imperfect, low level track-track maker correlations as both the track and bone records become better known. But as noted previously, ichnotaxonomy is (1) a parataxonomic scheme that aims to describe the morphology of the track independently, without the necessity of identifying the track maker, which may be unknown, and (2) a means of extracting useful information about both faunal diversity and paleoenvironment (and the relationships between the two) that is complementary to both body fossil-based paleontology and pure geology. Because of the consistent association of some track types with certain environments, the presence of those tracks provides information about a specific paleoenvironment that may otherwise be ambiguous based solely on the rocks themselves. Furthermore, tracks by themselves can provide unequivocal and intriguing evidence — an example is the sizable, terrestrial, zygodactyl bird that is unknown (and wholly unpredicted) from body fossils but that must be associated with the maker of the Early Cretaceous track *Shandongornipes* from China. Because limbs are generally lost early in the taphonomic decay process (Davis & Briggs, 1998; Brand et al., 2003), many fossil avian taxa based on skeletal remains completely lack feet (and, conversely, a few have been based solely on pedal elements). It is therefore conceivable, for example, that the *Shandongornipes* track maker could already be known from isolated, non-pedal elements, but in such circumstances it is impossible to correlate the remains with the footprints.

As noted in the sections that follow, careful description of track morphotypes and their sedimentological and paleoenvironmental contexts provides evidence that has a bearing on many aspects of avian distribution, behavior, ecology, and evolution. The track record thus provides an independent and complementary yardstick both for ichnotaxonomic and taxonomic evaluation of the fossil record.

Geographical and Geochronological Distribution Patterns

Assuming that different bird tracks are reasonably well-defined and differentiated in most cases, ichnotaxa can be used to understand track maker distribution in space and time. With the exception of two Southern Hemisphere reports of possible pre-Cretaceous bird tracks, all well-documented and well dated occurrences occur in the Cretaceous and Cenozoic. Thus, it appears that there was a pronounced avian radiation beginning close to the Jurassic-Cretaceous boundary about 145 million years ago. This independently tests and substantiates the avian body fossil record, with only *Archaeopteryx* (and possibly *Wellnhoferia* [Elzanowski, 2001], the validity of which is debatable [Mayr et al., 2007]) prior to the boundary but a much greater diversity of birds known from the Early Cretaceous. The anomalous, pre-Cretaceous occurrences include bird tracks from purportedly Upper Triassic strata of Argentina (Melchor et al., 2002; de Valais and Melchor, 2008) that were compared with *Gruipeda*. These deposits have not been dated with great confidence (Genise et al., 2008; L. Chiappe, personal communication). Other tracks, such as *Trisauropodiscus* from the Early Jurassic of southern Africa (Ellenberger, 1972,1974; Lockley et al., 1992), have also been attributed to birds, but they are more likely attributable to small, bipedal dinosaurs; some could even be assigned to the dinosaurian ichnogenus *Anomoepus* (Lockley and Gierliński, 2007).

The most notable aspect of Cretaceous bird track distributions is the high incidence of tracksite discovery in East Asia, notably in China and South Korea. This may be partly due to the abundance of lake deposits in these regions at this time. However, given that the contemporaneous Chinese and North Korean skeletal record is also rich in birds, the distribution patterns may reflect a true center of avian radiation — here again the track and body fossil records may be providing independent but complementary bodies of evidence substantiating the same hypothesis. Yet it is dangerous to draw anything other than tentative evolutionary conclusions without considering the incomplete and biased nature of the fossil record. The Cretaceous bird track record is poor in most other regions, although a modest record exists in North America, where the known diversity of bird tracks is considerably smaller, and it is tempting to infer that the avian faunas in separate regions were different. This inference is tentative, but supported to some degree by known endemicity among vertebrate faunas at certain times during the Cretaceous.

In striking contrast to the abundance of avian tracksites in the Cretaceous of East Asia, bird tracks are almost unknown in the Cretaceous of Europe. The situation is reversed in the Cenozoic: there are virtually no Cenozoic bird track sites reported from Asia, other than an Iranian Eocene (Mirzaie Ataabadi and Khazee, 2004) and Miocene (Abbassi and Shakeri, 2006) sites and a few Pliocene and Pleistocene sites in Iran, Japan, and Korea (Ono, 1984;

Okamura et al., 1993; Kim et al, 2009); only one Cenozoic bird track (*Iranipeda*) has ever been named from anywhere in Asia. But Cenozoic track sites are abundant in Europe, and their study has played a pivotal role in the development of avian ichnotaxonomy. (The tracks described and discussed by Kordos [1985] are now preserved and interpreted for the public in the Novohrad-Nógrád Geopark in northern Hungary.) Likewise, Cenozoic avian tracksites are common in the western United States, where European ichnotaxonomic conventions have generally been adopted with some modifications and amendments.

Many factors contribute to these regional and temporal differences in the avian track record, including the distributions of suitable sedimentary facies, the historical accident of discovery, and the priorities given to various ichnotaxonomic works in these different regions. At present, the bulk of the Mesozoic avian track record is based on sites and ichnotaxa described from South Korea and China, with a smaller contribution from North and South American sites and ichnotaxa. None are unequivocally known or convincingly described in detail from Europe, Africa, Australia, or Antarctica. In the case of Cenozoic track sites and ichnotaxa, the vast majority are known from Europe and North America, with a few reported from South America and Antarctica. Late Cenozoic bird tracks are also known from Australia (Rich and Green, 1974; Rich and Gill, 1976), New Zealand (Lockley et al., 2007a), Africa (Leakey and Hay, 1979), and east and central Asia, but as yet none of these have been formally named or attributed to existing ichnotaxa.

Behavioral Interpretations

Collective paleoecology and ichnofacies perspectives

As noted above, it is impossible to unambiguously infer the track makers of various bird track types, however distinctive they may be. Instead, descriptions use the well-understood foundation of extant birds as a basis. References to "shorebird-like," "duck-like," "heron-like,"or "roadrunner-like" tracks lack implications of certain association or relationship between the track makers and any extant taxon defined at a low taxonomic level (e.g., species or genus) — the descriptors merely provide a general basis for comparison. (The avian track and body fossil records are compared below.) Notwithstanding these constraints, specific behaviors on the parts of the track makers can be inferred directly from, and frequently associated with, particular ichnotaxa.

In a very general sense, many Mesozoic and Cenozoic sites with shorebird tracks are characterized by very high track densities (≥100 per m^2). Such high densities are typical of many modern lake margin settings to which true shorebirds flock to feed. Such track assemblages therefore indicate gregarious behavior and cannot parsimoniously be interpreted as the repeated activity of just a few individuals. Thus, gregariousness is strongly indicated in many shorebird-like ichnotaxa; that this behavior occurs even in Mesozoic ichnotaxa indicates its great antiquity: it may have simply been inherited from non-avian dinosaurs, many of which are also known to have exhibited gregarious behavior based on tracks.

The combination of evidence for predominantly gregarious shorebirds and their allies in lake margin settings gives rise to the shorebird ichnofacies concept (Lockley et al., 1994; Doyle et al. 2000). The definition of an ichnofacies is the recurrent association of a particular track (trace) type with a particular sedimentary facies representing a particular paleoenvironment (Seilacher, 1967; Bromley 1996). Since the shorebird ichnofacies was first defined (Lockley et al., 1994), with the implication that it could be traced back to the earliest recurrent assemblages of shorebird or shorebird-like tracks, many examples of the shorebird ichnofacies have been reported, spanning the Cretaceous through Cenozoic (Yang et al., 1995; Doyle, 2000; Lockley et al., 2006a, Kim et al., 2006; Lockley, 2007). Hunt and Lucas (2007, p. 66) even subsumed the shorebird ichnofacies into their archetypal *Grallator* ichnofacies, implying a correspondence between lake margin track assemblages with "tridactyl avian and non-avian theropods (usually dominant)." This, in turn, implies that small, early, non-avian theropod dinosaurs filled similar niches to later shorebirds. Hunt and Lucas (2007) explicitly defined two subdivisions (ichnocoenoses) of their *Grallator* ichnofacies: the Early Cretaceous *Jindongornipes* ichnocoenose and the Cenozoic *Avipeda* ichnocoenose (but they did not comment on the aforementioned ichnotaxonomic problems associated with the label *Avipeda*). This represents yet another possibly unnecessary and artificial distinction between Cretaceous and Cenozoic shorebird ichnotaxa and ichnofacies based on age, nomenclature, and taxonomic incongruence between Mesozoic and Cenozoic avifaunas (but not necessarily pedal morphologies and behaviors). Regardless, across all these studies one principle is clear: ichnofacies relationships persist through time as an expression of recurrent ecological conditions and the evolution of organisms to occupy niches in those ecosystems.

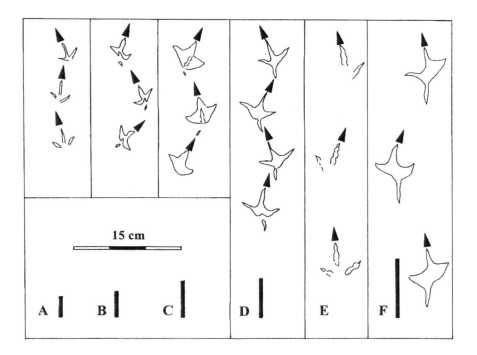

Figure 9. Shorebird-like trackways from the Mesozoic (Cretaceous) with black arrows indicating typical positively (inwardly) rotation. **A**: *Koreanaornis*; **B**: *Goseongornipes*; **C**: *Uhangrichnus*; **D**: *Ignotornis*; **E**: cf. *Jindongornipes*; **F**: cf. *Hwangsanipes*. All trackways, except *Ignotornis* from Colorado, are from the Cretaceous of Korea. Vertical black bars (**A-F**) correspond to track size (length), arranged in order of increasing size (except **E**, for which length is not measurable).

Individual behavior

Where individual trackways are recorded, many show a strong positive (inward) rotation, which is again typical of modern shorebirds, indicating either convergence or the antiquity of this trait (Figure 9). As indicated in previous sections, track size and morphology are often indistinguishable from those of modern bird tracks, indicating that through the evolution of birds, selective pressures and/or inherent morphodynamics (evolution of developmental programs) produced not only similar foot morphologies but even similar sizes and behaviors to birds in similar niches through time.

Of comparable interest are similarities in feeding behavior inferred from both tracks and bill traces. Three interesting examples can be cited. A few *Ignotornis* trackways demonstrate a "shuffling" behavior quite different from the normal walking pattern exhibited by other specimens (compare Figures 4 and 10). In these examples, the right and left feet each advanced for very short distances of less than a single footprint length. This has been interpreted as a type of "foot stirring" behavior characteristic of modern herons that "stir" up the substrate in search of food (Lockley et al., 2009). The holotype trackway of *Presbyornithiformipes*, the Eocene trackway of a duck- or flamingo-like bird shows a distinctive trail of "dabble marks" that are consistent in width with the bill of *Presbyornis* body fossils (Figure 6). These feeding trails show the slow back and forth motion of the beak relative to the trackway midline and the bird's direction of progression. Such behavior is typical of feeding waders. Perhaps the most striking example of feeding behavior was recently reported from the Cretaceous of Korea (Lockley et al., 2008b) where traces indistinguishable from those of modern spoonbills have been recorded (Figure 11). These traces consist of sets of fine, zigzag, arcuate to semicircular grooves that "sweep" back and forth across the trackway. These are indistinguishable from the feeding tracks of the modern spoonbill *Platalea* (Swennen and Yu, 2005), even though there is no fossil record of this group in the Cretaceous, or even any spoonbill morphology known in a Cretaceous bird. Here, the ichnological record permits the prediction that a Cretaceous bird with a spoonbill-like morphology will eventually be found, but even by itself the track record adds a "tic" to the list of known Cretaceous avian ecomorphologies and, therefore, diversity.

The Early Cretaceous, roadrunner-like trackway *Shandongornipes* is another example of the early appearance of an extant bird foot morphology and gait by a Mesozoic bird for which no body fossil record has yet been discovered. *Shandongornipes* tracks are zygodactyl and reminiscent in size, morphology, and gait of no bird other than the extant roadrunner, *Geococcyx*, which has no pre-Pleistocene fossil record. Again, the track record adds data to the body knowledge of Cretaceous avian diversity that the traditional body fossil record does not.

Collectively, the roadrunner-like footprints, Cretaceous species flocking and walking with shorebird-like gaits, and Cretaceous species exhibiting distinctively modern, heron- and spoonbill-like feeding behaviors may lead the reader, and even the ichnologist, to risk inferring that the track makers were essentially indistinguishable from modern species in foot morphology, gait and feeding behavior…and therefore possibly in phylogenetic relationships. ("If it walks like a duck…") However, the skeletal record suggests otherwise. Though the aim of ichnology is not necessarily to match up tracks with specific track makers, it is informative to compare the ichnological and body fossil records of birds through time to see where they complement or contradict one another.

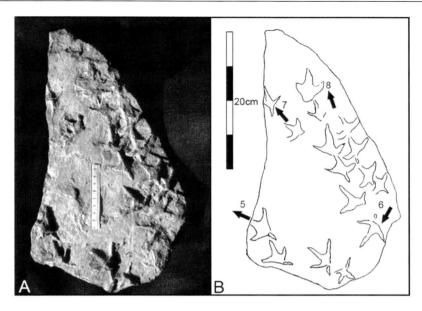

Figure 10. *Ignotornis* trackway demonstrating heron-like "foot stirring" behavior (after Lockley et al., 2009). Compare with Figure 4.

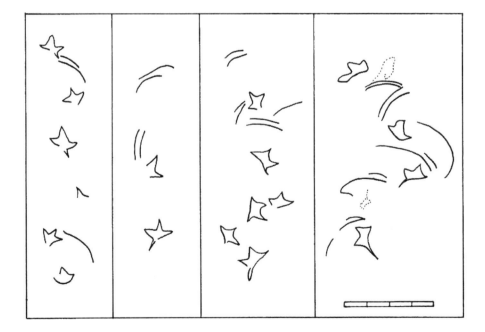

Figure 11. Arcuate, spoonbill-like feeding traces (A-D) associated with *Hwangsanipes*-like trackways (Figure 9F) from the Early Cretaceous of Korea are highly distinctive and indistinguishable from those of modern spoonbills. Access to these specimens, from the Gyeongsangnamdo Institute of Science Education, Korea was facilitated by the kindness of Dr. Jeong Yul Kim.

Consistency between the Avian Track and Bone Records

Hypothetically, there should be perfect consistency between the modern track record and the extant avifauna: i.e., every track-making bird can eventually be observed making tracks. Put another way, all bird tracks illustrated in reliable field guides have their "authors" identified unequivocally. But as every paleontology student learns, the similarity between extant and extinct species decreases as temporal distance from the present increases — older organisms are generally less like extant ones than are younger organisms. In fact, the different epochs of the Cenozoic Era, in order of increasing age (Holocene, Pleistocene, Pliocene, Miocene, Oligocene, Eocene and Paleocene), were originally defined on the basis of the decreasing proportions of extant taxa recognized by pioneer geologists (Lyell, 1830-1833). Therefore, the problems of determining track maker affinity increase with the increasing age of fossil footprints. As a result, it is far more difficult to infer the probable "authors" of Cretaceous tracks, even tracks very similar to those made by extant taxa, than those registered in the Pleistocene or Holocene.

For this reason, it is necessary to discuss what is known of the avian body fossil record in some detail before comparing and considering how the Mesozoic and Cenozoic bone records may correlate with their corresponding track records. The known body fossil record of Mesozoic birds has grown by at least an order of magnitude in just the past 25 years over what had been known in the previous 120 years (dating back to the first Mesozoic bird discoveries in the 1860s). Concurrent with this boom has been a fundamental shift in the understanding of avian evolution and ancestry from an ancestor of uncertain systematic position to a restricted clade of derived theropod dinosaurs (the Maniraptora). The body of literature documenting and supporting this theory is vast and beyond the scope of this paper (for excellent, recent reviews, see Witmer [2002] and Chiappe [2007]). For present purposes, it is sufficient to note that most of the long-standing "gaps" in 19th and 20th century understanding of avian evolution have been filled. The early perspective was based on very few data points because few Mesozoic birds — most notably *Archaeopteryx*, *Hesperornis*, and *Ichthyornis* — were known from material complete enough to provide relatively clear portraits of their "grades" with respect to modern birds. The filling of the substantial "gaps" separating extant (crown-group) birds (Neornithes *sensu* Padian et al. [1999] = Aves *sensu* Gauthier [1986]) from these few taxa, and from an amorphous archosaurian ancestry, has produced one of the best-documented evolutionary transitions in the vertebrate fossil record — so good that even a diagnosis of "bird" has become arbitrary and quite hazy (Chiappe and Witmer, 2002).

This revolution has revealed an early avian diversity of unsuspected richness (Figure 3). The bulk of these new discoveries have come from late Early Cretaceous (Barremian-Albian, ~128-110 Ma) localities in Spain (Sanz et al., 2002) and China (You et al., 2006; Zhou and Zhang, 2006a) and Late Cretaceous (Campanian-Maastrichtian, ~83-65 Ma) localities in Mongolia and Russia (Kurochkin, 2000), Argentina (Walker, 1981; Chiappe, 1996; Agnolin and Martinelli, 2008) and Antarctica (Noriega and Tambussi, 1995; Chatterjee, 2002; Clarke et al., 2005), though important specimens have also been found elsewhere and from other parts of the Cretaceous. Most famous are Spanish and Chinese Early Cretaceous specimens that come largely from lacustrine deposits, often preserving complete, articulated skeletons and frequently including soft tissues such as feathers, scales, webbing, and keratinous claw sheaths and beaks that provide unprecedented quantities of data about early birds.

Importantly, however, the vast majority of Mesozoic birds do not belong to Neornithes: most are not members of any modern, extant bird clade, and most are at best only distantly related to neornithians. The only Mesozoic bird fossils that have been attributed to neornithians are from the latest Cretaceous, and the identities and phylogenetic positions of most of these are controversial (Hope, 2002; Mayr, 2009). Therefore, most of the known Mesozoic record of birds — again, including more than half of avian evolutionary history — does not include neornithians. In terms of understanding the Mesozoic avian track record, this fact emphasizes the aforementioned danger in attributing tracks morphologically similar to those made by living birds to members of extant clades.

Archaeopteryx, from the latest Jurassic, has yet to be unseated as the most basal bird; most definitions of "bird" and "Aves" still hinge on this one taxon for traditional, though arbitrary, reasons (Chiappe and Witmer, 2002; Witmer, 2002; for alternative perspectives, in which *Archaeopteryx* and other taxa are birds but not avians, see Gauthier, [1986] and Gauthier and de Quieroz [2001]). It has since been joined at the base of the avian tree by other "long-tailed" birds such as the Late Jurassic *Wellnhoferia* (Elzanowski, 2001; q.v. Mayr et al., 2007) and Early Cretaceous *Jeholornis* (Zhou and Zhang, 2002, 2003b). The pedes of these birds do not appear anisodactyl (Zhou and Zhang, 2006a; Mayr et al., 2007), but neither are they plesiomorphically completely pamprodactyl. What kinds of tracks these birds may have made is unclear, depending on reconstructions of hallucal position and mobility.

Slightly more derived birds — pygostylians (Figure 3) — are characterized by the loss of the plesiomorphic long tail and the evolution of fully anisodactyl feet capable of perching. The most basal known pygostylians are late Early Cretaceous in age, exemplified by *Confuciusornis* (Hou et al., 1995a, b; Chiappe et al., 1999) and *Sapeornis* (Zhou and Zhang, 2001b, 2003a). Sometime in the Early Cretaceous, two lineages, the Enantiornithes (Walker, 1981) and the Ornithuromorpha (Chiappe, 2002b) (Euornithes of Sereno [1998]), split from within Pygostylia; together, these two clades constitute the smaller clade Ornithothoraces (Chiappe, 1996; Figure 3). Neornithes is a subclade nested deep within Ornithuromorpha, but based on the sheer number of fossils and taxa known, members of the now-extinct Enantiornithes were the most common Cretaceous birds.

Fossils described as avian but of uncertain phylogenetic positions began to be discovered in the early 1970s in Upper Cretaceous (Maastrichtian, ~70-65 Ma) strata of Argentina (Bonaparte and Powell, 1980) and Campanian (~83-70 Ma) strata of Mongolia (Elzanowski, 1976, 1977, 1981) and Mexico (Brodkorb, 1976). Other than *Archaeopteryx*, *Ichthyornis*, and a few hesperornithians, non-neornithian birds were essentially unknown when these discoveries were made, and they clearly did not much resemble the aforementioned Mesozoic birds. Comparisons were therefore made most strongly to various neornithians, and in fact some taxonomists attempted to shoehorn them into neornithian clades despite substantial morphological differences. The Argentine discoveries, however, led to the recognition of a new group of birds, the Enantiornithes (Walker, 1981; Chiappe and Walker, 2002), or "opposite birds" (so named because the peg-and-socket articulation between the scapula and coracoid in these birds is reversed from that of ornithuromorphs). Enantiornithians went extinct at the end of the Cretaceous but comprise the vast majority of Early Cretaceous ornithothoracian birds and substantial portions of many Late Cretaceous avifaunas. Most Early Cretaceous taxa were conservative in pedal morphology: anisodactyl with long halluces. They have been generally perceived as adapted for perching (Hopson, 2001; Chiappe and Walker, 2002; Sereno et al., 2002; but see Glen and Bennett, 2007) and would be expected to have registered tracks resembling those made by extant columbids or most passerines. This is true even for longirostrine enantiornithians like *Longipteryx*, *Longirostravis*, and *Shanweiniao* whose rostra appear adapted for charadriiform-like probe feeding or piscivory (Zhang et al., 2001; Hou et al., 2004; O'Connor et al., 2009). Some Late Cretaceous (Maastrichtian) enantiornithians, such as *Yungavolucris*, *Lectavis*, and *Neuquenornis*, demonstrate a greater diversity of pedal morphologies (Chiappe, 1993; Chiappe and Calvo, 1994; Chiappe and Walker, 2002) that presumably correlate with a broader range of paleoecologies, but since many of these taxa are known only from isolated tarsometatarsi, potential track morphologies cannot be inferred from them. Because enantiornithians are the most common Cretaceous birds, it is tempting to infer that members of this clade are most likely to be the makers of Cretaceous bird tracks.

Compared to their enantiornithian sister taxa, ornithuromorphs are much rarer in the Cretaceous, particularly the Early Cretaceous. An early possible ornithuromorph, *Liaoningornis* (Hou et al., 1996; Hou, 1997), from the late Early Cretaceous of China, was anisodactyl with a large hallux, similar to those of "perching" enantiornitheans. Other, contemporaneous basal ornithuromorphs, such as *Archaeorhynchus* (Zhou and Zhang, 2006b), *Hongshanornis* (Zhou and Zhang, 2006c), and *Yanornis* and *Yixianornis* (Zhou and Zhang, 2001a), however, had far shorter halluces, as well as pedal phalangeal proportions and ungual morphologies, suggestive of more terrestrial lifestyles (Hopson, 2001). The skull morphology and diet of *Yanornis* suggest it may have been partly aquatic (Zhou et al., 2004). The most basal known ornithuromorph, *Patagopteryx*, is a flightless bird from the mid Late Cretaceous of Argentina that has secondarily pamprodactyl feet (Chiappe, 2002a).

More derived ornithuromorphs belong in the clade Ornithurae (*sensu* Chiappe, 2002b; Figure 3). The oldest and most basal known ornithuran, *Gansus*, from the late Early Cretaceous of China, was originally interpreted as charadriiform-like based on pedal proportions (Hou and Liu, 1984) but is now known to have had anseriform-like webbed feet, with the webbing extending as far as the bases of the pedal unguals (You et al., 2006). All other ornithurans are Late Cretaceous in age (the hesperornithian *Enaliornis* from England may just predate the Early-Late Cretaceous boundary, ~100 million years ago [Galton and Martin, 2002]). The basal ornithuran *Apsaravis*, from the Campanian of Mongolia, has terrestrial pedal phalangeal proportions and may have lost digit I (Clarke and Norell, 2002). More derived are the most famous Cretaceous ornithurans: the flightless, diving hesperornithians, such as *Hesperornis* and *Baptornis*, which had webbed, pamprodactyl feet. The pelvic morphologies of these birds suggests that they were poorly adapted for terrestrial locomotion, so tracks made by hesperornithians may be very rare and could be predicted to exhibit unique gaits and stances, possibly convergent with modern loons or even penguins. The equally famous and coeval *Ichthyornis*, as well as *Apatornis* and *Iaceornis* (Clarke, 2004), are the closest known sister taxa to Neornithes; they plus Neornithes constitute the clade Carinatae. Feet are unknown in these basal carinates, but they are often reconstructed as procellariiform-like and may have had similar pedal morphologies.

Neornithes — modern birds — were long perceived as a wholly Cenozoic phenomenon because no Mesozoic fossils could inarguably be attributed to any neornithian clade, though many fragmentary fossils were (and are still) often placed in neornithian clades (e.g., Brodkorb, 1963, 1969; Cracraft, 1972; Olson and Parris, 1987; Hope, 1999; Olson, 1999; see Hope [2002] for a thorough review of all Mesozoic fossils attributed at one time or another to neornithian clades). More recently, unambiguous, latest Cretaceous members of Anseriformes have been discovered (e.g., Kurochkin et al., 2002; Clarke et al., 2005). The basal position of Anseriformes in Neornithes (Figure 3) enables the prediction that Cretaceous members of the Galliformes (sister taxon of Anseriformes within the Galloanserae), as well as paleognaths (phylogenetically the most basal neornithians), must also have been present, and indeed potential candidates have been reported (Parris and Hope, 2002; Varricchio, 2002; Agnolin et al., 2003; Clarke, 2004). More interestingly, possible latest Cretaceous neoavians (all neornithians except paleognaths and galloanserans; Figure 3) have also been reported, including Pelecaniformes, Charadriiformes (Hope, 2002), Gaviiformes (Lambrecht, 1929; Olson, 1992; Chatterjee, 2002), and, most ambiguously, Psittaciformes (Stidham, 1998, 1999; q.v. Dyke and Mayr, 1999). More complete fossils are necessary to support these assignments, however.

This detailed excursion into the Mesozoic avian body fossil record is essential if we are to understand the significance of the coeval avian track record. The body fossil record strongly indicates that, except in the latest Cretaceous, there were no modern birds (Neornithes) in the Mesozoic, despite many molecular phylogenetic studies that imply the presence of at least stem neornithians as far back as the Early Cretaceous (e.g., Brown et al., 2008). That many Mesozoic footprints strongly resemble those made by neornithians can therefore be explained only by one of two hypotheses: (1) pre-latest Cretaceous neornithians were indeed present and made many of the Mesozoic tracks, but the coeval body fossil record is, for unknown reasons, strongly biased toward the preservation of non-neornithian birds, or (2) many non-neornithians converged with later neornithians in foot morphology (or, more correctly, later neornithians converged with their non-neornithian, Mesozoic predecessors), and non-neornithians were responsible for registering the Mesozoic tracks. While the first option is possible — early neornithians may have lived in upland, non-depositional areas, for example — it requires too many caveats and is far less parsimonious than the second, which is therefore the preferred (and only empirical) hypothesis. Given this, the further implication is that evolutionary convergence, at least in pedal morphologies and ecological niches, was rampant between Cenozoic and Mesozoic birds.

No tracks have been attributed to the most basal, "long-tailed" birds. The absence of anisodactyly, plus the brevity of the hallux in these taxa, suggests that they would have registered tracks indistinguishable from those made by non-avian theropod dinosaurs. Tracks made by these taxa, therefore, may be known but perceived as small, non-avian theropod tracks. Distinguishing between small, gracile, non-avian theropod (or other tridactyl dinosaur) tracks is a difficult issue to resolve, particularly in cases of tracks made by small, maniraptoran theropods, many of which have foot morphologies that would register tracks indistinguishable from those of basal avians (though some have distinct pedal morphologies and left easily recognizable tracks [Li et al., 2007]). The similarities of many small theropod pedes to those of birds has led to the identification of many, especially pre-Early Cretaceous, tracks as avian despite the both real and predicted absence of birds in those time periods (see Lockley and Rainforth [2002] for a review). This discrepancy continues even through the present with the aforementioned *Gruipeda*-like tracks reported from Argentina.

Most striking, however, is that many, perhaps most, Mesozoic avian footprints have no known body fossil counterparts, particularly in the Early Cretaceous. As above, the most common Early Cretaceous birds — enantiornithians and more basal taxa — conservatively have anisodactyl, possibly perching feet, but the most common Early Cretaceous and early Late Cretaceous bird footprints have "shorebird-like" or "wading bird-like," often tridactyl, morphologies that are more suggestive of the pedal morphologies of rare, Early Cretaceous, basal ornithuromorphs (with the exception of *Shandongornipes*, for which no known Mesozoic bird has matching pedal morphology). The only potentially good match between an Early Cretaceous avian taxon and a contemporaneous ichnotaxon is the basal ornithuran *Gansus*: its webbed, functionally tridactyl feet match fairly well webbed *Uhangrichnus* tracks. Web-footed tracks are also known from the Late Cretaceous (*Hwangsanipes*, *Ignotornis*, and *Sarjeantopodus*), as well as the Cenozoic, but *Gansus* is not known beyond the late Early Cretaceous. Moreover, there has not yet been any study of whether or not *Gansus* digital proportions match those of *Uhangrichnus* tracks. It is unlikely that *Gansus* was the only web-footed bird in the Early Cretaceous — indeed, some Cretaceous web-footed tracks possess elongate hallux impressions, but *Gansus* lacks a long hallux — so it cannot be implicated as the definitive track maker of coeval web-footed tracks.

Only neornithians are known to have survived the end-Cretaceous (K-P$_g$) extinction event; the post-extinction diversification appears to have been relatively rapid for neoavians because virtually all major clades have at least basal representatives by the Middle Eocene, and many even in the Paleocene. The Cenozoic bird track record, beginning in the earliest Paleocene (Johnson, 1986), is dominated by tracks that resemble those made by extant birds, beginning the Cenozoic trend in which the track record much more closely matches the body fossil record than does the Mesozoic track-body fossil dichotomy. Yet the danger remains of drawing artificial distinctions between Mesozoic and Cenozoic bird tracks simply because the contemporaneous avifaunas of the two eras are so distinct: the Mesozoic and Cenozoic bird track records are much more similar to one another than are the avifaunas from each time period. Again, this underscores the importance of the track record: it provides the unique perspective that the relative constancy between the Mesozoic and Cenozoic track records reflects the convergent evolution of adaptive morphologies to particular lifestyles in specific ecosystems, without a need to match tracks to members of particular clades.

Evolutionary Significance

Generally speaking, there is broad agreement between the track and body fossil records regarding the origination and extinction of major vertebrate groups. For example, the first tetrapod body fossils more or less coincide with the first appearance of their tracks. Likewise, there is close correspondence between the origination and extinction of dinosaur body fossils and their track record. However, when examined in detail, the track record can add significantly to our understanding of some of the more subtle aspects of evolution.

The track record of Mesozoic birds is instructive in this regard, suggesting on the surface the presence of a range of abundant shorebird-like neornithians at a time when the corresponding skeletal record of such modern forms is nonexistent. The reasons for this disparity may ultimately pertain to extrinsic, geological and environmental factors that control preservation, rather than intrinsic, biological dynamics that may also have a strong influence on evolution. The Mesozoic track record may be an example of a system biased toward the preservation of tracks made by taxa whose paleoecologies put them in track-making (and track-preserving) situations more often than others. In the Mesozoic, these may have been the rarer, possibly aquatic ornithuromorphs rather than the more common, arboreal enantiornithians, but regardless of the actual affinities of these track makers, they are not currently known from body fossils. In recognizing these disparities, as well as correspondences, we gain valuable insight into the completeness of the fossil record and the limits of reasonable inference. The Mesozoic bird footprint record is therefore both exceptionally informative about and complementary to the coeval skeletal fossil record.

Tracks and body fossils are often preferentially preserved in different environments, and the correspondence between the two records may vary from very high to zero depending on the preservation potential of the rock units being studied (Lockley and Hunt, 1994). Thus, reported Late Triassic tracks from Argentina attributed to a bird with pigeon-like feet (Melchor et al., 2002) have potentially far reaching implications, especially if the reported age is proven well-founded (see Genise et al. [2008] for comment). But to date, no equivalent body fossils are known, and claims of avian skeletons pre-dating the 150 million year old *Archaeopteryx* have proved controversial and ambiguous at best; none have yet withstood close scrutiny. Although the Argentine tracks may well be avian in origin, more evidence is needed to verify the age determination, and until such evidence is widely accepted, it is difficult to draw conclusions about bird origins from such isolated reports.

However, by the Early Cretaceous, the weight of evidence for an important avian radiation is far more compelling, and it is tempting to locate it in East Asia. In South Korea, many Cretaceous formations are replete with multiple layers that yield bird tracks even though there is no Korean skeletal record corresponding to the dominant footprint morphologies, and with the possible exception of *Gansus*, none elsewhere, either. But bird fossils are well-known in other parts of East Asia, including North Korea (Li and Gao, 2007) and China (references above). In China, the coeval bird track record is more scattered and there is little correlation with the skeletal record, even though in some cases both records derive from broadly similar lake basin deposits. Nevertheless, when evidence is pooled, the region is clearly important for both tracks and body fossils. The situation is similar in the Cretaceous of North America, where there is little correspondence between the skeletal and track records for birds: the track records occur in terrestrial and marginal marine deposits of late Early, early Late, and late Late Cretaceous age and made by birds with terrestrial foot morphologies, but the body fossil record is thus far dominated by fossils of aquatic birds from mid-Late Cretaceous marine deposits. Again the most complete picture derives from a pooling of data: one data set complements the other to provide a greater picture of Cretaceous North American avifaunas than either does alone.

Such evidence provides a cautionary tale by highlighting gaps in the skeletal record. It is instructive to contrast the disparate avian track-skeletal records in the Cretaceous with the far more highly correlated records in the Cenozoic. In other words, the adage about decreasing confidence of interpreting faunas with the "lens of the present" with time holds true in the avian record. Without resorting to speculation, is impossible to infer the proportions of taxa not represented due to incompleteness of the fossil record, but the avian track record at least provides evidence for a diversity of Cretaceous bird types (morphologies) not represented in the skeletal record, and this contrasts with the much closer correlations evident in comparing the Cenozoic track and skeletal records.

CONCLUSIONS

1. Historically, the study of fossil bird footprints is inextricably linked with the study of avian and non-avian dinosaur osteological remains. The fact that bird and non-avian theropod dinosaur tracks were confused for half a century prior to the 1860s underscores the now well-established close relationship between these groups.

2. Because ornithologists may be unfamiliar with the conventions and nomenclature of paleoichnology, we stress that ichnotaxonomy is a parataxonomic system and that track-track maker correlations are not necessarily the primary aim of paleoichnology. The determination of the "author" of a given fossil footprint at the "family," genus or species level may be possible only in rare cases.

3. The growing paleoichnological literature has a significant impact on vertebrate paleobiology and paleoenvironmental analysis. At present, the literature on avian fossil footprints comprises ~200 papers that describe at least 38 avian footprint ichnogenera comprising ~65 ichnospecies. Of these, 15 and 23 ichnogenera, and 20 and 45 ichnospecies, have been named from the Mesozoic and Cenozoic, respectively.

4. Traditionally, Mesozoic and Cenozoic tracks have been treated and named separately — the lack of taxonomic overlap is partly a reflection of artificial specializations and distinctions. Differences between some ichnotaxa are subtle and some duplication or taxonomic splitting is evident and in need of further study.

5. In the Cenozoic, bird tracks are more easily grouped into morphofamiles (ichnofamilies) and ichnogenera, such as *Gruipeda*, *Ardeipeda*, and *Charadriipeda*, originally defined on the basis of relatively young (Miocene) assemblages described from Europe.

6. Although many ichnotaxa based on Cenozoic tracks may correspond to extant avian families (if not genera or species), attempts to correlate tracks with potential track makers are inherently speculative. Potential correlations between fossil tracks and extant taxa become increasingly difficult as one moves back in time and are particularly problematic in the Mesozoic, when the avifauna, as understood from the body fossil record, was substantially different from Cenozoic avifaunas.

7. The vast majority of avian ichnotaxa represent shorebird, wader, and other water bird species (or, in the Mesozoic, taxa like these in ecology and pedal morphology) that were active in lake and ocean margin settings. Avian diversity inferred in such settings is comparable to modern settings.

8. The second most common category of avian fossil footprints is that of large, ground dwelling species like the moa, emu, and rhea. Such tracks are primarily Cenozoic; few such bird tracks are known from the Mesozoic.

9. In the Early Cretaceous, avian ichnotaxa, especially from east Asia, show remarkable morphological and behavioral convergence with extant neornithian taxa (e.g., roadrunners, ducks, herons, spoonbills), despite a lack of corresponding skeletal evidence for such taxonomic groups and scarcely any evidence of taxa with morphologies matching the tracks.

10. Current evidence suggests that during the Early Cretaceous, birds were particularly abundant and diverse in the regions that are now East Asia. This may indicate that

the area was a center of for an avian evolutionary radiation. However, differences in the preservation potential between this and other regions may also contribute to this apparent pattern.

REFERENCES

Abbassi, N. & Shakeri, S. (2006). Miocene vertebrate footprints from the Upper Red Formation, Muchampa area, Zanjan Province. `Ulum-i zamin [Geosciences], 55*, 76-89.

Agnolin, F. L. & Martinelli, A. G. (2008). Fossil birds from the Late Cretaceous Los Alamitos Formation, Río Negro Province, Argentina. *Journal of South American Earth Sciences, 27*, 42-49.

Agnolin, F. L. Novas, F. E. & Lio, G. (2003). Restos de un posible galliforme (Aves, Neornithes) del Cretácico Tardio de Patagonia. *Ameghiniana, 40*, 49R.

Alonso, R. N. Carabajal, E. & Raskovsky, M. (1980). Hallazgo de icnitas (Aves, Charadriformes) en el Terciario de la Puna Argentina. *Actas del Segundo Congreso Argentino de Paleontología y Bioestratigrafía y Primer Congreso Latinoamericano de Paleontología, Buenos Aires, 1978, 3*, 75-83.

Alonso, R. N. & Marquillas, R. A. (1986). Nueva localidad con huellas de dinosaurios y primer hallazgo de huellas de Aves en la Formacion Yacoraite (Maastrichtiano) del norte Argentino. *Actas IV Congreso Argentino de Paleontología y Bioestrategrafía, Mendoza, 2*, 33-41.

Anfinson, O. A. Lockley, M. G. Kim, S. H. Kim, K. S. & Kim, J. Y. (2009). First report of the small bird track *Koreanaornis* from the Cretaceous of North America: implications for avian ichnotaxonomy and paleoecology. *Cretaceous Research, 30*, 885-894.

Aramayo, S. (2009). A brief sketch of the Monte Hermosa human footprint site, South coast of Buenos Aires Province, Argentina. *Ichnos, 16*, 49-54.

Aramayo, S. A. & Manera de Bianco, T. (1987). Hallazgo de una icnofauna continental (Pleistoceno tardio) en la localidad de Pehuen-Co (Partido de Coronel Rosales) Provincia de Buenos Aires, Argentina. Parte I: Edentata, Litopterna, Proboscidea. Parte II: Carnivora, Artiodactyla y Aves. *IV Congreso Latinoamericano de Paleontologia Actas, 1*, 516-547.

Azuma, Y. Tomida, Y. & Currie, P. J. (2002). Early Cretaceous bird tracks from the Tetori Group, Fukui Prefecture, Japan. *Memoir of the Fukui Prefectural Dinosaur Museum, 1*, 1-6.

Barthel, K. W. Swinburne, N. H. M. & Morris, S. C. (1990). *Solnhofen: A Study in Mesozoic Palaeontology*. Cambridge University Press, Cambridge, 236 pp.

Bonaparte, J. F. & Powell, J. E. (1980). A continental assemblage of tetrapods from the Upper Cretaceous beds of El Brete, northwestern Argentina (Sauropoda-Coelurosauria-Carnosauria-Aves); 19-28 in Taquet, P. (Ed.) *Ecosystèmes Continentaux du Mésozoïque. Mémoires de la Société Géologique de France, N.S.* 139.

Brand, L. R. Hussey, M. & Taylor, J. (2003). Decay and disarticulation of small vertebrates in controlled experiments. *Journal of Taphonomy, 1*, 69-95.

Brodkorb, P. (1963). Birds from the Upper Cretaceous of Wyoming; 55-70 in Sibley, C. G. (Ed.) *Proceedings of the 13th Ornithological Congress*. The Amerian Ornithologists' Union, Baton Rouge.

Brodkorb, P. (1969). The generic position of a Cretaceous bird. *Quarterly Journal of the Florida Academy of Sciences, 32*, 239-240.

Brodkorb, P. (1976). Discovery of a Cretaceous bird, apparently ancestral to the Orders Coraciiformes and Piciformes (Aves: Carinatae); 67-73 in Olson, S. L. (Ed.) *Collected Papers in Avian Paleontology Honoring the 90th Birthday of Alexander Wetmore. Smithsonian Contributions to Paleobiology, 27*.

Bromley, R. G. (1996). *Trace Fossils: Biology, Taphonomy and Applications, 2nd Ed.* Chapman & Hall, London, 361 pp.

Brown, J. W. Rest, J. S. Garcia-Moreno, J. Sorenson, M. D. & Mindell, D. P. (2008). Strong mitochondrial DNA support for a Cretaceous origin of modern avian lineages. *BMC Biology, 6*, 6 (1-18).

Brustur, T. (1997). The stages of the paleoichnological studies in Romania. *Geo-Eco-Marina, 2*, 205-216.

Buffetaut, E. (2004). Footprints of giant birds from the Upper Eocene of the Paris Basin: an ichnological enigma; 357-362 in Pemberton, S. G. McCrea, R. T. & Lockley, M. G. (Eds.) *William Antony Swithin Sarjeant (1935-2002): A Celebration of His Life and Ichnological Contributions, Volume 3. Ichnos, 11*.

Chatterjee, S. (2002). The morphology and systematics of *Polarornis*, a Cretaceous loon (Aves: Gaviidae) from Antarctica; 125-155 in Zhou, Z. & Zhang, F. (Eds.) *Proceedings of the 5th Symposium of the Society of Avian Paleontology and Evolution, Beijing, 1-4 June 2000*. Science Press, Beijing.

Chiappe, L. M. (1993). Enantiornithine (Aves) tarsometatarsi from the Cretaceous Lecho Formation of northwestern Argentina. *American Museum Novitates* 3083, 1-27.

Chiappe, L. M. (1996). Late Cretaceous birds of southern South America: anatomy and systematics of Enantiornithes and *Patagopteryx deferrariisi*. *Münchner Geowissenschaften Abhandlungen, 30*, 203-244.

Chiappe, L. M. (2002a). Osteology of the flightless *Patagopteryx deferrariisi* from the Late Cretaceous of Patagonia (Argentina); 281-316 in Chiappe, L. M. & Witmer, L. M. (Eds.) *Mesozoic Birds: Above the Heads of Dinosaurs*. University of California Press, Berkeley.

Chiappe, L. M. (2002b). Basal bird phylogeny: problems and solutions; 448-472 in Chiappe, L. M. & Witmer, L. M. (Eds.) *Mesozoic Birds: Above the Heads of Dinosaurs*. University of California Press, Berkeley.

Chiappe, L. M. (2007). *Glorified Dinosaurs: the Origin and Early Evolution of Birds*. Wiley-Liss, Hoboken, 192 pp.

Chiappe, L. M. & Calvo, J. O. (1994). *Neuquenornis volans*, a new Late Cretaceous bird (Enantiornithes: Avisauridae) from Patagonia, Argentina. *Journal of Vertebrate Paleontology, 14*, 230-246.

Chiappe, L. M. & Walker, C. A. (2002). Skeletal morphology and systematics of the Cretaceous Euenantiornithes (Ornithothoraces: Enantiornithes); 240-267 in Chiappe, L. M. & Witmer, L. M. (Eds.) *Mesozoic Birds: Above the Heads of Dinosaurs*. University of California Press, Berkeley.

Chiappe, L. M. & Witmer, L. M. (2002). Preface; ix-xii in in Chiappe, L. M. & Witmer, L. M. (Eds.) *Mesozoic Birds: Above the Heads of Dinosaurs*. University of California Press, Berkeley.

Chiappe, L. M. Ji, S. Ji, Q. & Norell, M. A. (1999). Anatomy and systematics of the Confuciusornithidae (Theropoda: Aves) from the Late Mesozoic of northeastern China. *Bulletin of the American Museum of Natural History, 242*, 1-89.

Clarke, J. A. (2004). Morphology, phylogenetic taxonomy, and systematics of *Ichthyornis* and *Apatornis* (Avialae: Ornithurae). *Bulletin of the American Museum of Natural History, 286*, 1-179.

Clarke, J. A. & Norell, M. A. (2002). The morphology and phylogenetic position of *Apsaravis ukhaana* from the Late Cretaceous of Mongolia. *American Museum Novitates, 3387*, 1-46.

Clarke, J. A. Tambussi, C. P. Noriega, J. I. Erickson, G. M. & Ketcham, R. A. (2005). Definitive fossil evidence for the extant avian radiation in the Cretaceous. *Nature, 433*, 305-309.

Cockburn-Hood, T. H. (1874). Remarked upon the footprints of the *Dinornis* in the sand rock at Poverty Bay, New Zealand, and upon its recent extinction. *Proceedings of the Royal Society of Edinburgh, 8*, 236-240.

Cope, E. D. (1876). On a gigantic bird from the Eocene of New Mexico. *Proceedings of the Academy of Natural Sciences of Philadelphia, 28(2)*, 10-11.

Coria, R. A. Currie, P. J. Eberth, D. & Garrido, A. (2002). Bird footprints from the Anacleto Formation (Late Cretaceous), Neuquén, Argentina. *Ameghiniana, 39*, 453-463.

Covacevich, R. & Lamperein, C. (1970). Hallazgo de icnitas en Peninsula Fildes, Isla Rey Jorge, Archipelago Setland del Sur, Antarctica. *Series Cientifico del Instituto Antartido Chileno, 1*, 55-74.

Covacevich, R. & Lamperein, C. (1972). Ichnites from Fildes Peninsula, King George Island, South Shetland Islands. *Antarctic Geology and Geophysics, IUGS Series B, 1*, 71-74.

Cracraft, J. (1972). A new Cretaceous charadriiform family. *Auk, 89*, 36-46.

Currie, P. J. (1981). Bird footprints from the Gething Formation (Aptian, Lower Cretaceous) of northeastern British Columbia, Canada. *Journal of Vertebrate Paleontology, 1*, 257-264.

de Raaf, J. F. M. (1965). Lower Oligocene bird-tracks from northern Spain. *Nature, 207*, 146-148.

Darwin, C. R. (1859). *On the Origin of Species by Means of Natural Selection, or the Preservation of Favoured Races in the Struggle for Life*. John Murray, London.

Davis, P. G. (1997). The bioerosion of bird bones. *International Journal of Osteoarchaeology, 7*, 388-401.

Davis, P. G. & Briggs, D. E. G. (1998). The impact of decay and disarticulation on the preservation of fossil birds. *Palaios, 13*, 3-13.

Delair, J. B. (1989). A history of dinosaur footprint discoveries in the British Wealden; 19-25 in Gillette, D. D. & Lockley, M. G. (Eds.) *Dinosaur Tracks and Traces*. Cambridge University Press, Cambridge.

Desnoyers, J. (1859). Note sur des empreintes de pas d'animaux dans le gypse des environs de Paris, particulièrement de la vallée de Montmorency. *Comptes Rendus Hebdomadaires des Séances de l'Academie des Sciences (Paris), 49*, 67-73.

de Valais, S. & Melchor, R. N. (2008). Ichnotaxonomy of bird-like footprints: an example from the Late Triassic-Early Jurassic of northwest Argentina. *Journal of Vertebrate Paleontology, 28*, 145-159.

Doyle, P. Wood, J. L. & George, G. T. (2000). The shorebird ichnofacies: an example from the Miocene of southern Spain. *Geological Magazine, 137*, 517-536.

Dyke, G. J. & Mayr, G. (1999). Did parrots exist in the Cretaceous period? *Nature, 399*, 317-318.

Elbroch, M. & Marks, E. (2001). *Bird Tracks & Sign: A Guide to North American Species.* Stackpole Books, Mechanicsburg, 456 pp.

Ellenberger, P. (1972). Contribution a la classification des pistes de vertebres du Trias: les types du Stormberg d'Afrique du Sud (I). *Palaeovertebrata Memoire Extraordinaire*, 1-152.

Ellenberger, P. (1974). Contribution a la classification des pistes de vertebres du Trias: les types du Stormberg d'Afrique du Sud (II). *Palaeovertebrata Memoire Extraordinaire*, 1-202.

Ellenberger, P. (1980). Sur les empreintes de pas de gros mammifères de l'Éocene supérieur de Garrigues Ste Eulalie (Gard). *Palaeovertebrata Memoire Jubilaire de Rene Lavocat*, 37-77.

Elzanowski, A. (1976). Paleognathous bird from the Cretaceous of central Asia. *Nature, 264*, 51-53.

Elzanowski, A. (1977). Skulls of *Gobipteryx* (Aves) from the Upper Cretaceous of Mongolia. *Palaeontologica Polonica, 37*, 153-165.

Elzanowski, A. (1981). Embryonic bird skeletons from the Late Cretaceous of Mongolia. *Palaeontologica Polonica, 42*, 147-179.

Elzanowski, A. (2001). A new genus and species for the largest specimen of *Archaeopteryx*. *Acta Palaeontologica Polonica, 46*, 519-532.

Erickson, B. R. (1967). Fossil bird tracks from Utah. *Museum Observer, 5*, 140-146.

Fornós, J. J. Bromley, R. G. Clemmensen, L. B. & Rodríguez-Perea, A. (2002). Tracks and trackways of *Myotragus balearicus* Bate (Artiodactyla, Caprinae) in Pleistocene aeolianites from Mallorca (Balearic Islands, western Mediterranean). *Palaeogeography, Palaeoclimatology, Palaeoecology, 180*, 277-313.

Fuentes Vidarte, C. (1996). Primeras huellas de Aves en el Weald de Soria (España). Nuevo icnogenero, *Archaeornithipus* y nueva icnoespecie *A. meijidei*. *Estudios Geologicos, 52*, 63-75.

Galton, P. M. & Martin, L. D. (2002). *Enaliornis*, an Early Cretaceous hesperornithiform bird from England, with comments on other Hesperornithiformes; 317-338 in Chiappe, L. M. & Witmer, L. M. (Eds.) *Mesozoic Birds: Above the Heads of Dinosaurs*. University of California Press, Berkeley.

Gauthier, J. (1986). Saurischian monophyly and the origin of birds; 1-55 in Padian, K. (Ed.) *The Origin of Birds and the Evolution of Flight. Memoirs of the California Academy of Sciences, 8*.

Gauthier, J. & de Quieroz, K. (2001). Feathered dinosaurs, flying dinosaurs, crown dinosaurs, and the name "Aves"; 7-41 in Gauthier, J. & Gall, L. F. (Eds.) *New Perspectives on the Origin and Early Evolution of Birds*. Peabody Museum of Natural History, New Haven.

Genise, J. F. Melchor, R. N. Archangelsky, M. Bala, L. O. Straneck, R. & de Valais, S. (2008). Application of neoichnological studies to behavioural and taphonomic interpretation of fossil bird-like tracks from lacustrine settings: the Late Triassic-Early Jurassic? Santo Domingo Formation, Argentina. *Palaeogeography, Palaeoclimatology, Palaeoecology, 272*, 143-161.

Geoffroy Saint Hilaire, É. (1851). Première partie. Mammifères. Catalogue des primates, in *Catalogue Méthodique de la Collection des Mammifères de la Collection des Oiseaux et des Collections Annexés*. Muséum d'Histoire Naturelle de Paris. Gide et Baudry, Paris, 96 pp.

Gillies, T. B. (1872). On the occurrence of footprints of the moa at Poverty Bay. *Transactions and Proceedings of the New Zealand Institute, 4*, 127-128.

Glen, C. L. & Bennett, M. B. (2007). Foraging modes of Mesozoic birds and non-avian theropods. *Current Biology, 17*, R911-R912.

Greben, R. & Lockley, M. G. (1993). The Mesozoic and Cenozoic bird track record: a bias in favor of shore-birds. *Journal of Vertebrate Paleontology* 13(suppl. 3), 38A-39A.

Grozescu, H. (1918). Geologié de la région subcarpatique de la parte septentrionale du district de Bacău. *Extrait de l'Annuaire de l'Institut Geologique de Roumanie, 8*, 1-46.

Hill, H. (1895). On the occurrence of moa-footprints in the bed of the Manawatu River, near Palmerston North. *Transactions and Proceedings of the Royal Society of New Zealand, 27*, 476-477.

Hill, H. (1913). The moa-legendary, historical, and geological: why and when the moa disappeared. *Transactions and Proceedings of the Royal Society of New Zealand, 46*, 330-351.

Hitchcock, E. (1836). Ornithichnology. — Description of the foot marks of birds, (*Ornithichnites*) on new Red Sandstone in Massachusetts. *American Journal of Science, 29*, 307-340.

Hitchcock, E. (1858). *Ichnology of New England: a Report on the Sandstone of the Connecticut Valley*, Especially its Fossil Footmarks. William White, Boston, 199 pp.

Hitchcock, E. (1861). Remarks upon certain points in ichnology. *Proceedings of the American Association for the Advancement of Science, 14*, 144-156.

Hope, S. (1999). A new species of *Graculavus* from the Cretaceous of Wyoming (Aves: Neornithes); 261-266 in Olson, S. L. Wellnhofer, P. Mourer-Chauviré, C. Steadman, D. W. & Martin, L. D. (Eds.) *Avian Paleontology at the Close of the 20th Century: Proceedings of the 4th International Meeting of the Society of Avian Paleontology and Evolution, Washington,* D.C. 4-7 June 1996. *Smithsonian Contributions to Paleobiology, 89.*

Hope, S. (2002). The Mesozoic radiation of Neornithes; 339-388 in Chiappe, L. M. & Witmer, L. M. (Eds.) *Mesozoic Birds: Above the Heads of Dinosaurs*. University of California Press, Berkeley.

Hopson, J. A. (2001). Ecomorphology of avian and nonavian theropod phalangeal proportions: implications for the arboreal versus terrestrial origin of bird flight; 211-235 in Gauthier, J. & Gall, L. F. (Eds.) *New Perspectives on the Origin and Early Evolution of Birds*. Peabody Museum of Natural History, New Haven.

Hou, L. (1997). A carinate bird from the Upper Jurassic of western Liaoning, China. *Chinese Science Bulletin, 42*, 413-416.

Hou, L. & Liu, Z. (1984). A new fossil bird from Lower Cretaceous of Gansu and early evolution of birds. *Scientia Sinica, Series B, 27*, 1296-1301.

Hou, L. Zhou, Z. Gu, Y. & Zhang, H. (1995a). *Confuciusornis sanctus*, a new Late Jurassic sauriurine bird from China. *Chinese Science Bulletin, 40*, 1545-1551.

Hou, L. Zhou, Z. Martin, L. D. & Feduccia, A. (1995b). A beaked bird from the Jurassic of China. *Nature, 377*, 616-618.

Hou, L. Martin, L. D. Zhou, Z. & Feduccia, A. (1996). Early adaptive radiation of birds: evidence from fossils from northeastern China. *Science, 274*, 1164-1167.

Hou, L. Chiappe, L. M. Zhang, F. & Chuong, C. M. (2004). New Early Cretaceous fossil from China documents a novel trophic specialization for Mesozoic birds. *Naturwissenschaften, 91*, 22-25.

Hunt, A. P. & Kelley, S. A. (2004). Bird footprints from the Miocene 'Abiquiu' Formation of north-central New Mexico with a review of Cenozoic tetrapod tracks of New Mexico. *New Mexico Geology, 26*, 65.

Hunt, A. P. & Lucas, S.G. (2007). Tetrapod ichnofacies: a new paradigm. *Ichnos, 14*, 59-68.

Huxley, T. H. (1867). On the classification of birds; and on the taxonomic value of the modification of certain of the cranial bones observable in that class. *Proceedings of the Scientific Meetings of the Zoological Society of London, 1867*, 415-472.

Huxley, T. H. (1868). On the animals which are most nearly intermediate between birds and reptiles. *Geological Magazine, 5*, 357-365

Huxley, T. H. (1870). Further evidence of the affinity between the dinosaurian reptiles and birds. *Quarterly Journal of the Geological Society of London, 26*, 12-31.

Johnson, K. R. (1986). Paleocene bird and amphibian tracks from the Fort Union Formation, Bighorn Basin, Wyoming. *University of Wyoming Contributions to Geology, 24*, 1-10.

Kim, B. K. (1969). A study of several sole marks in the Haman Formation. *Journal of the Geological Society of Korea, 5*, 243-258.

Kim, J. Y. Kim, S. H. Kim, K. S. & Lockley, M. G. (2006). The oldest record of webbed bird and pterosaur tracks from South Korea (Cretaceous Haman Formation, Changseon and Sinsu Islands): more evidence of high avian diversity in east Asia. *Cretaceous Research, 26*, 56-69.

Kim, S. K. Kim, J. Y. Kim, S. H. Lee, C. Z. & Lim, J. D. (2009). Preliminary report on hominid and other vertebrate footprints from the Late Quaternary strata of Jeju Island, Korea. *Ichnos, 16*, 1-11.

Kordos, L. (1985). Footprints in Lower Miocene sandstone at Ipolytarnóc, N. Hungary. *Geologica Hungarica, Series Palaeontologica, 46*, 257-415.

Kordos, L. & Prakfalvi, P. (1990). A contribution to the knowledge of Neogene beds with footprints marks in Europe. *Annual Report of the Geological Institute of Hungary, 1988*, 201-212.

Krapovickas, V. Ciccolii, P. L. Managano, G. & Limarino, C. O. (2007). Vertebrate and invertebrate trace fossils in anastamosing fluvial deposits of the Toro Negro Formation (Upper Miocene) Argentina, 51-52 in Lucas, S. G. Spielmann, J. A. & Lockley, M. G. (Eds.) *Cenozoic Vertebrate Tracks and Traces. New Mexico Museum of Natural History and Science Bulletin, 42*.

Kurochkin, E. N. (2000). Mesozoic birds of Mongolia and the former USSR; 533-559 in Benton, M. J. Shishkin, M. A. Unwin, D. M. & Kurochkin, E. N. (Eds.) *The Age of Dinosaurs in Russia and Mongolia*. Cambridge University Press, Cambridge.

Kurochkin, E. N. Dyke, G. J. & Karhu, A. A. (2002). A new presbyornithid bird (Aves, Anseriformes) from the Late Cretaceous of southern Mongolia. *American Museum Novitates, 3386*, 1-11.

Lambrecht, K. (1929). *Neogaeornis wetzeli* n. g. n. sp. der erste Kreidevogel der südlichen Hemisphäre. *Paläontologische Zeitschrift, 11*, 121-129.

Lambrecht, K. (1938). *Urmiornis abeli* n. sp. eine Pliozäne Vogelfährte aus Persian. *Palaeobiologica, 6*, 242-245.

Leakey, M. D. & Hay, R. L. (1979). Pliocene footprints in the Laetolil Beds at Laetoli, northern Tanzania. *Nature, 278*, 317-323.

Lee, Y. N. (1997). Bird and dinosaur footprints in the Woodbine Formation (Cenomanian), Texas. *Cretaceous Research, 18*, 849-864.

Leonardi, G. (1994). *Annotated Atlas of South America Tetrapod Footprints (Devonian to Holocene)*. Companhia de Pesquisa de Recursos Minerais, Rio de Janeiro, 247 pp.

Li, Q. & Gao, K. Q. (2007). Lower Cretaceous vertebrate fauna from the Sinuiju basin, North Korea as evidence of geographic extension of the Jehol Biota into the Korean Peninsula. *Journal of Vertebrate Paleontology* 27(suppl. 3), 106A.

Li, R. Lockley, M. G. & Liu, M. (2005). A new ichnotaxon of fossil bird track from the Early Cretaceous Tianjialou Formation (Barremian-Albian), Shandong Province, China. *Chinese Science Bulletin, 50*, 1149-1154.

Li, R. Lockley, M. G. Makovicky, P. J. Matsukawa, M. Norell, M. A. Harris, J. D. & Liu, M. (2007). Behavioral and faunal implications of Early Cretaceous deinonychosaur trackways from China. *Naturwissenschaften, 95*, 185-191.

Livezey, B. C. & Zusi, R. L. (2007a). Higher-order phylogeny of modern birds (Theropoda, Aves: Neornithes) based on comparative anatomy: I. Methods and characters. *Bulletin of the Carnegie Museum of Natural History, 37*, 1-544.

Livezey, B. C. & Zusi, R. L. (2007b). Higher-order phylogeny of modern birds (Theropoda, Aves: Neornithes) based on comparative anatomy: II. — analysis and prospects. *Zoological Journal of the Linnean Society, 149*, 1-95.

Lockley, M. G. (1998). The vertebrate track record. *Nature, 396*, 429-432.

Lockley, M. G. (2007). A tale of two ichnologies: the different goals and missions of vertebrate and invertebrate ichnology and how they relate in ichnofacies analysis. *Ichnos, 14*, 39-57.

Lockley, M. G. Chin, K. Houck, M. Matsukawa, M. & Kukihara, R. (2009). New interpretations of *Ignotornis* the first-reported Mesozoic avian footprints: implications for the ecology and behavior of an enigmatic Cretaceous bird. *Cretaceous Research, 30*, 1041-1061.

Lockley, M. G. & Delgado, C. R. (2007). Tracking an ancient turkey: a preliminary report on a new Miocene ichnofauna from near Durango, Mexico; 67-72 in Lucas, S. G. Spielmann, J. A. & Lockley, M. G. (Eds.) *Cenozoic Vertebrate Tracks and Traces. New Mexico Museum of Natural History and Science Bulletin, 42*.

Lockley, M. G. & Gierliński, G. (2006). Diverse vertebrate ichnofaunas containing *Anomoepus* and other unusual trace fossils from the Lower Jurassic of the western United States: implications for paleoecology and palichnostratigraphy; 175-191 in Harris, J. D. Lucas, S. G. Spielmann, J. A. Lockley, M. G. Milner, A. R. C. & Kirkland, J. I. *The Triassic-Jurassic Terrestrial Transition. New Mexico Museum of Natural History and Science Bulletin, 37*.

Lockley, M. G. Gregory, M. R. & Gill, B. (2007a). The ichnological record of New Zealand's moas: a preliminary summary; in 73-78 in Lucas, S. G. Spielmann, J. A. & Lockley, M. G. (Eds.) *Cenozoic Vertebrate Tracks and Traces. New Mexico Museum of Natural History and Science Bulletin, 42.*

Lockley, M. Harris, J. Chin, K. & Matsukawa, M. (2008b). Ichnological evidence for morphological and behavioral convergence between Mesozoic and extant birds: paleobiological implications; 71 in Uchman, A. (Ed.) *Second International Congress on Ichnology Abstract Book.* Polish Geological Institute, Warszawa.

Lockley, M. G. Houck, K. Yang, S. Y. Matsukawa, M. & Lim, S. K. (2006a). Dinosaur dominated footprint assemblages from the Cretaceous Jindong Formation, Hallayo Haesang National Park, Goseong County, South Korea: evidence and implications. *Cretaceous Research, 26*, 70-101.

Lockley, M. G. & Hunt, A. P. (1994). A review of vertebrate ichnofaunas of the Western Interior United States: evidence and implications; 95-108 in Caputo, M. V. Peterson, J. A. & Franczyk, K. J. (Eds.) *Mesozoic Systems of the Rocky Mountain Region, United States.* Rocky Mountain Section, Society of Economic Paleontologists and Mineralogists, Denver.

Lockley, M. G. Hunt, A. P. & Meyer, C. (1994). Vertebrate tracks and the ichnofacies concept: implications for paleoecology and palichnostratigraphy; 241-268 in Donovan, S. (Ed.) *The Paleobiology of Trace Fossils.* Wiley and Sons, Inc. New York.

Lockley, M. G. Kim, J. Y. Kim, K. S. Kim, S. H. Matsukawa, M. Li, R. Li, J. & Yang, S. Y. (2008a). *Minisauripus* – the track of a diminutive dinosaur from the Cretaceous of China and South Korea: implications for stratigraphic correlation and theropod foot morphodynamics. *Cretaceous Research, 29*, 115-130.

Lockley, M. G. Li, R. Harris, J. D. Matsukawa, M. & Liu, M. (2007b). Earliest zygodactyl bird feet: evidence from Early Cretaceous roadrunner-like tracks. *Naturwissenschaften, 94*, 657-665.

Lockley, M. Matsukawa, M. Ohira, H. Li, J. Wright, J. White, D. & Chen, P. (2006a). Bird tracks from Liaoning Province, China: new insights into avian evolution during the Jurassic-Cretaceous transition. *Cretaceous Research, 27*, 33-43

Lockley, M. G. & Meyer, C. A. (2000). *Dinosaur Tracks and Other Fossil Footprints of Europe.* Columbia University Press, New York, 323.

Lockley, M. G. Nadon, G. & Currie, P. J. (2004). A diverse dinosaur-bird footprint assemblage from the Lance Formation, Upper Cretaceous, eastern Wyoming: implications for ichnotaxonomy; 229-249 in Pemberton, S. G. McCrea, R. T. & Lockley, M. G. (Eds.) *William Antony Swithin Sarjeant (1935-2002): A Celebration of His Life and Ichnological Contributions, Volume 3. Ichnos, 11.*

Lockley, M. G. & Rainforth, E. C. (2002). The track record of Mesozoic birds and pterosaurs: an ichnological and paleoecological perspective; 405-418 in Chiappe, L. M. & Witmer, L. M. (Eds.) *Mesozoic Birds: Above the Heads of Dinosaurs.* University of California Press, Berkeley.

Lockley, M. G. Reynolds, R. E. Milner, A. R. C. & Varhalmi, G. (2007c). Preliminary overview of mammal and bird tracks from the White Narrows Formation, Southern Nevada; 91-96 in Lucas, S. G. Spielmann, J. A. & Lockley, M. G. (Eds.) *Cenozoic Vertebrate Tracks and Traces. New Mexico Museum of Natural History and Science Bulletin, 42.*

Lucas, S. G. (2007). Cenozoic vertebrate footprints named by O. S. Vyalov in 1965 and 1966, 113-148 in Lucas, S. G. Spielmann, J. A. & Lockley, M. G. (Eds.) *Cenozoic Vertebrate Tracks and Traces*. *New Mexico Museum of Natural History and Science Bulletin, 42*.

Lucas, S. G. Kelley, S. A. Spielmann, J. A. Lockley, M. G. & Connell, S. D. (2007). Miocene Bird footprints from Northern New Mexico, 169-176 in Lucas, S. G. Spielmann, J. A. & Lockley, M. G. (Eds.) *Cenozoic Vertebrate Tracks and Traces*. *New Mexico Museum of Natural History and Science Bulletin, 42*.

Lucas, S. G. & Schultz, G. E. (2007). Miocene vertebrate footprints from the Texas panhandle, 177-183 in Lucas, S. G. Spielmann, J. A. & Lockley, M. G. (Eds.) *Cenozoic Vertebrate Tracks and Traces*. *New Mexico Museum of Natural History and Science Bulletin, 42*.

Lockley, M. G. Wright, J. L. & Matsukawa, M. (2001). A new look at *Magnoavipes* and so-called "Big Bird" tracks from Dinosaur Ridge (Cretaceous, Colorado); 137-146 in Lockley M. G. & Taylor, A. (Eds.) *Dinosaur Ridge: Celebrating a Decade of Discovery. Mountain Geologist, 38*, 137-146.

Lockley, M. G. Yang, S. Y. Matsukawa, M. Fleming, F. & Lim, S. K. (1992). The track record of Mesozoic birds: evidence and implications. *Philosophical Transactions of the Royal Society of London B, 336*, 113-134.

Lull, R. S. (1904). Footprints of the Jura-Trias of North America. *Memoir of the Boston Society of Natural History, 5*, 461-557.

Lull, R. S. (1953). Triassic life of the Connecticut Valley. *State Geological and Natural History Survey of Connecticut Bulletin, 81*, 1-333.

Lyell, C. (1830-33). *Principles of Geology Vols. 1-3*. John Murray, London.

Mayr, G. (2009). *Paleogene Fossil Birds*. Springer-Verlag, Berlin, 262 pp.

Mayr, G. Pohl, B. Hartman, S. & Peters, D. S. (2007). The tenth skeletal specimen of *Archaeopteryx*. *Zoological Journal of the Linnean Society, 149*, 97-116.

McCrea, R. T. & Sarjeant, W. A. S. (2001). New ichnotaxa of bird and mammal footprints from the Lower Cretaceous (Albian) Gates Formation of Alberta; 453-478 in Tanke, D. H. & Carpenter, K. (Eds.) *Mesozoic Vertebrate Life*. Indiana University Press, Bloomington.

Mehl, M. G. (1931). Additions to the vertebrate record of the Dakota Sandstone. *American Journal of Science, 21*, 441-452.

Melchor, R. (2009). Bird tracks preserved in fluvial channel facies of the Rio Negro Formation (Neogene), La Pampa Province Argentina. *Ameghiniana, 46*, 209-214.

Melchor, R. N. de Valais, S. & Genise, G. F. (2002). Bird-like fossil footprints from the Late Triassic. *Nature, 417*, 936-938.

Meunier, S. (1906). Remarquables traces de pas sur un banc de gypse. *Le Naturaliste, 28(453)*, 19-21.

Milàn, J. (2006). Variations in the morphology of emu (*Dromaius novaehollandiae*) tracks reflecting differences in walking pattern and substrate consistency: ichnotaxonomic implications. *Palaeontology, 49*, 405-420.

Mirzaie Ataabadi, M. & Khazaee, A. R. (2004). New Eocene mammal and bird footprints from Birjand area, eastern Iran; 363-370 in Pemberton, S.G. McCrea, R. T. & Lockley, M. G. (Eds.) *William Antony Swithin Sarjeant (1935-2002): A Celebration of His Life and Ichnological Contributions, Volume 3. Ichnos, 11*.

Morgan, G. S. & Williamson, T. E. (2007). Middle Miocene (Late Barstovian) mammal and bird tracks from the Benavidez Ranch localfauna, Zia Fomation, Albuquerque Basin, Sandoval County, New Mexico, 319-330 in Lucas, S. G. Spielmann, J. A. & Lockley, M. G. (Eds.) *Cenozoic Vertebrate Tracks and Traces. New Mexico Museum of Natural History and Science Bulletin, 42.*

Murelaga, X. Astibia, H. Baceta, J. I. Almar, Y. Beamud, B. & Larrasoaña, J. C. (2007). Fósiles de pisadas de aves en el Oligoceno de Etaio (Navarra, Cuenca del Ebro). *Geogaceta, 41*, 139-142.

Mustoe, G. E. (2002). Eocene bird, reptile, and mammal tracks from the Chuckanut Formation, northwest Washington. *Palaios, 17*, 403-413.

Noriega, J. J. & Tambussi, C. P. (1995). A Late Cretaceous Presbyornithidae (Aves: Anseriformes) from Vega Island, Antarctic Peninsula: paleobiogeographic implications. *Ameghiniana, 32*, 57-61.

O'Connor, J. K. Wang, X. Chiappe, L. M. Gao,C. Meng, Q. Cheng, X. & Liu, J. (2009). Phylogenetic support for a specialized clade of Cretaceous enantiornithine birds with information from a new species. *Journal of Vertebrate Paleontology, 29*, 188-204.

Okamura, Y. Takehashi, K. & Lake Biwa Museum Research Associates. (1993). Fossil footprints of a bird from the Kobiwako Group, central Japan. *Fossils, 55*, 9-15.

Olson, S. L. (1992). *Neogaeornis wetzeli* Lambrecht, a Cretaceous loon from Chile (Aves: Gaviidae). *Journal of Vertebrate Paleontology, 12*, 122-124.

Olson, S. L. (1999). The anseriform relationships of *Anatalavis* Olson and Parris (Anseranatidae), with a new species from the Lower Eocene London Clay; 231-243 in Olson, S. L. Wellnhofer, P. Mourer-Chauviré, C. Steadman, D. W. & Martin, L. D. (Eds.) Avian Paleontology at the Close of the 20th Century: Proceedings of the 4th International Meeting of the Society of Avian Paleontology and Evolution, Washington, D.C. 4-7 June 1996. *Smithsonian Contributions to Paleobiology, 89.*

Olson, S. L. & Parris, D. C. (1987). The Cretaceous birds of New Jersey. *Smithsonian Contributions to Paleobiology, 63*, 1-23.

Ono, K. (1984). Fossil wading birds from northeast Honshu, Japan. *Memoirs of the National Science Museum of Tokyo, 17*, 39-46.

Owen, R. (1842a). Notice of the fragment of a femur of a gigantic bird of New Zealand. *Transactions of the Zoological Society of London, 3*, 29-33.

Owen, R. (1842b). Report on British fossil reptiles. Part II. *Report of the British Association for the Advancement of Science, 1841*, 60-204.

Owen, R. (1879). Memoir on the *Ornithichnites*, or foot-prints of species of *Dinornis*; 451-453 in Owen, R. (Ed.) *Memoirs on the Extinct Wingless Birds of New Zealand;* with an Appendix on those of England, Australia, Newfoundland, Mauritius, and Rodriguez (2 vols). Van Voorst, London.

Padian, K. Hutchinson, J. R. & Holtz, T. R. Jr. (1999). Phylogenetic definitions and nomenclature of the major taxonomic categories of the carnivorous Dinosauria (Theropoda). *Journal of Vertebrate Paleontology, 19*, 69-80.

Panin, N. (1965). Coexistence de traces de pas de vertébrés et de mécanoglyphes dans le Molasse Miocene des Carpates orientales. *Revue Roumaine de Géologie, Géophysique, et Géographie, 7*, 141-163.

Panin, N. & Avram, E. (1962). Noi orme de vertebrate în Miocenul Subcarpaţilor Romîneşti. *Studii şi Cercetari de Geologie, Geofizica, Geografie, 7*, 455-484.

Parris, D. C. & Hope, S. (2002). New interpretations of the birds from the Navesink and Hornerstown formations, New Jersey, USA (Aves: Neornithes); 113-124 in Zhou, Z. & Zhang, F. (Eds.) *Proceedings of the 5th Symposium of the Society of Avian Paleontology and Evolution,* Beijing, 1-4 June 2000. Science Press, Beijing.

Patterson, J. & Lockley, M. G. (2004). A probable *Diatryma* track from the Eocene of Washington: an intriguing case of controversy and skepticism; 341-347 in Pemberton, S. G. McCrea, R. T. & Lockley, M. G. (Eds.) *William Antony Swithin Sarjeant (1935-2002): A Celebration of His Life and Ichnological Contributions, Volume 3. Ichnos, 11.*

Payros, A. Astiba, H. Cearreta, A. Pereda Suberbiola, X. Murelaga, X. & Badiola, A. (2000). The Upper Eocene South Pyrenean coastal deposits (Liedena Sandstone, Navarra). Sedimentary facies, benthic foraminifera and avian ichnology. *Facies, 42,* 107-132.

Peabody, F. E. (1955). Taxonomy and the footprints of tetrapods. *Journal of Paleontology, 29,* 915-924.

Portis, A. (1879). Intorno ad alcune impronte eoceniche di vertebrati recentemente scoperte in Piemonte. *Atti della Reale Accademia delle Scienze, 15,* 221-229.

Remika, P. (1999). Identification, stratigraphy, and age of Neogene vertebrate footprints from the Vallecito-Fish Creek Basin, Anza-Borrego Desert State Park, California. *San Bernardino County Museum Association Quarterly, 46(2),* 37-45.

Rich, P. V. & Gill, E, (1976). Possible dromornithid footprints from Pleistocene dune sands of southern Victoria, Australia. *Emu, 76,* 221-223.

Rich, P. V. & Green, R. H. (1974). Footprints of birds at south Mt. Cameron, Tasmania. *Emu, 74,* 245-248.

Sarjeant, W. A. S. (1989). "Ten Paleoichnological Commandments": a standardized procedure for the description of fossil vertebrate footprints; 369-370 in Gillette, D. D. & Lockley, M. G. (Eds.) *Dinosaur Tracks and Traces.* Cambridge University Press, Cambridge.

Sarjeant, W. A. S. (1990). A name for the trace of an act: approaches to the nomenclature and classification of fossil footprints; 299-307 in Carpenter, K. & Currie, P. J. (Eds.) *Dinosaur Systematics: Perspectives and Approaches.* Cambridge University Press, Cambridge.

Sarjeant, W. A. S. & Langston, W. (1994). Vertebrate footprints and invertebrate traces from the Chadronian (Late Eocene) of Trans-Pecos Texas. *Bulletin of Texas Memorial Museum, 36,* 1-86.

Sarjeant, W. A. S. & Reynolds, R. E. (2001). Bird footprints from the Miocene of California, 21-40 in Reynolds, R. E. (Ed.) *The Changing Face of the East Mojave Desert.* California State University Fullerton, Desert Studies Consortium, Fullerton.

Sanz, J. L. Pérez-Moreno, B. P. Chiappe, L. M. & Buscalioni, A. D. (2002). The birds from the Lower Cretaceous of Las Hoyas (province of Cuenca, Spain); 209-229 in Chiappe, L. M. & Witmer, L. M. (Eds.) *Mesozoic Birds: Above the Heads of Dinosaurs.* University of California Press, Berkeley.

Seilacher, A. (1967). The bathymetry of trace fossils. *Marine Geology, 5,* 413-428.

Sereno, P. C. (1998). A rationale for phylogenetic definitions, with application to the higher-level taxonomy of Dinosauria. *Neues Jahrbuch für Geologie und Paläontologie Abhandlungen, 210,* 41-83.

Sereno, P. C. Rao, C. & Li, J. (2002). *Sinornis santensis* (Aves: Enantior-nithes) from the Early Cretaceous of northeastern China; 184-208 in Chiappe, L. M. & Witmer, L. M.

(Eds.) *Mesozoic Birds: Above the Heads of Dinosaurs*. University of California Press, Berkeley.

Stidham, T. (1998). A lower jaw from a Cretaceous parrot. *Nature, 396*, 29-30.

Stidham, T. A. (1999). Did parrots exist in the Cretaceous period? *Nature, 399*, 318.

Swennen, C. & Yu, Y. T. (2005). Food and feeding behavior of the black faced spoonbill. *Waterbirds, 28*, 19-27.

Switek, B, in press. T. H. Huxley's reptile-faced birds and bird-legged dinosaurs; in Moody, R. (Ed.) *Dinosaurs — A Historical Perspective. Geological Society of London Special Publication*.

Thode, S. (2008). Bones and words in 1870s New Zealand: the moa-hunter debate through actor networks. *British Journal for the History of Science, 42*, 225-244.

Thrasher, L. (2007). Fossil trackways from the Bowie Zeolite mines, Graham County Arizona, 269-273 in Lucas, S. G. Spielmann, J. A. & Lockley, M. G. (Eds.) *Cenozoic Vertebrate Tracks and Traces. New Mexico Museum of Natural History and Science Bulletin, 42*.

Trapani, J. (1998). Hydrodynamic sorting of avian skeletal remains. *Journal of Archaeological Science, 25*, 477-487.

Varricchio, D. J. (2002). A new bird from the Upper Cretaceous Two Medicine Formation of Montana. *Canadian Journal of Earth Sciences, 39*, 19-26.

Vialov, O. S. (1965). *Stratigrafiya Neogenovix Molass Predcarpatskogo Progiba [Neogene Stratigraphy of the Ciscarpathian Basin Molasse]*. Naukova Dumka, Kiev, 192 pp. (in Russian).

Vialov, O. S. (1966). *Sledy Zhiznedeiatel'nosti Organizmow I ikh Paleontogicheskoe Znachenie [Traces of the Activity of Organisms and their Paleontological Meaning]*. Naukova Dumka, Kiev, 219 pp. (in Russian).

Vialov, O. S. (1989). Pliocene bird tracks from Iran assigned to the genus Urmiornis. *Palaeontological Journal, 23*, 119-121.

Voy, C. D. (1880). On the occurrence of footprints of *Dinornis* at Poverty Bay, New Zealand. *American Naturalist, 14*, 682-684.

Williams, W. L. (1872). On the occurrence of footprints of a large bird, found at Turanganui, Poverty Bay. *Transactions and Proceedings of the Royal Society of New Zealand, 4*, 124-127.

Walker, C. A. (1981). New subclass of birds from the Cretaceous of South America. *Nature, 292*, 51-53.

Witmer, L. M. (2002). The debate on avian ancestry: phylogeny, function, and fossils; 3-30 in Chiappe, L. M. & Witmer, L. M. (Eds.) *Mesozoic Birds: Above the Heads of Dinosaurs*. University of California Press, Berkeley.

Yang, S. Y. Lockley, M. G. Greben, R. Erickson, B. R. & Lim, S. K. (1995). Flamingo and duck-like bird tracks from the Late Cretaceous and Early Tertiary: evidence and implication. *Ichnos, 4*, 21-34.

You, H. L. Lamanna, M. C. Harris, J. D. Chiappe, L. M. O'Connor, J. Ji, S. A. Lü, J. C. Yuan, C. X. Li, L. D. Zhang, X. Lacovara, K. J. Dodson, P. & Ji, Q. (2006). A nearly modern amphibious bird from the Early Cretaceous of northwestern China. *Science, 312*, 1640-1643.

Zhang, F. Zhou, Z. Hou, L. & Gu, G. (2001). Early diversification of birds: evidence from a new opposite bird. *Chinese Science Bulletin, 46*, 945-949.

Zhen, S. Li, J. Chen, W. & Zhu, S. (1995). Dinosaur and bird footprints from the Lower Cretaceous of Emei County, Sichuan. *Memoirs of the Beijing Natural History Museum, 54*, 105-120.

Zhou, Z. Clarke, J. Zhang, F. & Wings, O. (2004). Gastroliths in *Yanornis*: an indication of the earliest radical diet-switching and gizzard plasticity in the lineage leading to living birds. *Naturwissenschaften, 91*, 571-574.

Zhou, Z. & Zhang, F. (2001a). Two new ornithurine birds from the Early Cretaceous of western Liaoning, China. *Chinese Science Bulletin, 46*, 1258-1264.

Zhou, Z. & Zhang, F. (2001b). Largest bird from the Early Cretaceous and its implications for the earliest avian ecological diversification. *Naturwissenschaften, 89*, 34-38.

Zhou, Z. & Zhang, F. (2002). A long-tailed, seed-eating bird from the Early Cretaceous of China. *Nature, 418*, 405-409.

Zhou, Z. & Zhang. F. (2003a). Anatomy of the primitive bird *Sapeornis chaoyangensis* from the Early Cretaceous of Liaoning, China. *Canadian Journal of Earth Sciences, 40*, 731-747.

Zhou, Z. & Zhang, F. (2003b). *Jeholornis* compared to *Archaeopteryx*, with a new understanding of the earliest avian evolution. *Naturwissenschaften, 90*, 220-225.

Zhou, Z. H. & Zhang, F. C. (2006a). Mesozoic birds of China — a synoptic review. *Vertebrata PalAsiatica, 44*, 74-98.

Zhou, Z. & Zhang, F. (2006b). A beaked basal ornithurine bird (Aves, Ornithurae) from the Lower Cretaceous of China. *Zoologica Scripta, 35*, 363-373.

Zhou, Z. & Zhang, F. (2006c). Discovery of an ornithurine bird and its implication for Early Cretaceous avian radiation. *Proceedings of the National Academy of Sciences, 102*, 18998-19002.

In: Trends in Ornithology Research
Editors: P. K. Ulrich and J. H. Willett, pp. 49-94

ISBN: 978-1-60876-454-9
© 2010 Nova Science Publishers, Inc.

Chapter 2

TROPHIC RELATIONSHIPS AND MECHANISMS OF ECOLOGICAL SEGREGATION AMONG HERON SPECIES IN THE PARANA RIVER FLOODPLAIN (BIRDS: ARDEIDAE)

Adolfo H. Beltzer[1], Juan A. Schnack[2], Martín A. Quiroga[1], María de la Paz Ducommun[1] , Ana Laura Ronchi Virgolini[1,3] and Viviana Alessio[4]

[1]Instituto Nacional de Limnología (INALI, CONICET - UNL), Paraje El Pozo s/n, Ciudad Universitaria, 3000 Santa Fe, Argentina
[2]Facultad de Ciencias Naturales y Museo de La Plata (U.N.L.P.), Paseo del Bosque s/n, (1900) La Plata , Buenos Aires, Argentina
[3]Centro de Investigación Científica y Tecnológica con Transferencia a la Producción (CONICET)
[4]Facultad de Ciencia y Tecnología - Universidad Autónoma de Entre Ríos, 3100, Paraná Andrés Pazos y Corrientes

ABSTRACT

Herons are one of the best represented families in the floodplain of the Paraná River. The fact that interspecific competition constitutes the most significant factor in resources distribution is a prevailing idea in the ecological theory. According to recent studies, even though competition is important, the modeling of the community's structure results from the combined action of other factors which operate independently from the interspecific interaction. The distribution of resources is closely related to the ecological niche concept, this being the quantitative description of the organic unit requirements. It is hypothesized in our work that the studied species: *Ardea cocoi* (White-necked Heron), *Butotides striatus* (Striated Heron), *Bubulcus ibis* (Cattle Egret), *Ardea_alba (*Great Egret), *Egretta thula* (Snowy Egret), Syrigma sibilatrix (Whistling Heron), *Nycticorax nycticorax* (Black-crowned Night Heron), *Tigrisoma lineatum* (Fasciated Tiger-Heron) and *Ixobrychus involucris* (Stripe-backed Bittern), despite of the observed sympatry, have developed adaptative mechanisms of ecological segregation. This allows these species to use the resources in such a way that their diet composition (trophic sub niche) and other parameters of their ecological requirements (temporal sub niche and spatial sub niche) are differentiated. The index of relative importance was applied to calculate the contribution of each food category to the diet of each species. The trophic overlapping of the alimentary spectra, accumulated trophic diversity, alimentary efficiency, dietary selectivity, trophic spread of the niche, spatial sub niche and habitat preference were estimated. As regards trophic spectrum, even though fishes were found to be the basic diet for all four species and insects the second food category, slight differences exist between them which would establish mechanisms at the catches level. This is reinforced by the low overlapping values obtained and the lack of significance in the selectivity values. Variations concerning temporal and spatial sub niches were also obtained. Summarizing, the coexistence is mainly based on the differential utilization of the resources as basic isolation mechanisms and less subtly on space and time. Without leaving aside the usefulness of further research, we think these results provide valuable data for the understanding of the Paraná complex system dynamics.

INTRODUCTION

Wetlands exhibit a high concentration of wildlife. Their productivity is usually higher than that of terrestrial systems, which allows remarkable concentrations of fauna not found in other environments (Canevari *et al.*, 1999).

In general, birds constitute a characteristic component of worldwide aquatic systems and they may be considered indicators of the state of water bodies, their productivity at different trophic levels and the peculiarities of their structure and function. They are outstanding consumers within this type of systems (Martinez, 1993), and play an important role in the transfer of energy from these systems to terrestrial ones. Birds can obtain their food from different environmental units of the aquatic system by means of the spatial differential use of the environment or the ecospace (Dobzhansky *et al.*, 1983).

Even though, according to Maitland (1978) and Martínez (1993), only a few birds can be regarded as completely aquatic. Approximately 8 out of the 28 current orders present morphological and physiological adaptations related to living in aquatic environments (Ziswiler, 1980). Within their checklist, it can be identified from species just feeding on fish but which present no remarkable adaptations to aquatic life such as the Kingfisher (Beltzer and Oliveros, 1987) and the Osprey (Beltzer *et al.*, 2001), to those species which, like the Horned Grebe, build floating nests and hardly ever leave the water (Beltzer, 1983; Beltzer and Oliveros, 1982).

Many typically aquatic birds are originally terrestrial, which have adjusted to life in the water by means of a likely unfinished process during which they have developed morphological, physiological and ethological adaptations.

Herons (Ardeidae), non-diving aquatic birds, have changed their appearance to a considerably lesser extent. The main external changes are expressed through the relative lengthening of legs, beak, and their great flying capacity, which has enabled them to explore diverse permanent and temporal water bodies. The fresh-water medium provides resources to many species of very different lineages using such environments in various ways or feeding strategies.

Many studies on the trophic ecology of herons, their association with habitats, food resources and feeding spectrum have been reported (Kushlan, 1976a, 1976b, 1978, 1981; Amat and Soriguer, 1981; Amat, 1984; McNeil *et al.*, 1993; Lekuona and Campos, 1995). Many ardeid species are regarded as the most important components of the bird community associated with different environments of the middle Paraná River (Beltzer, 1981, 2003, Beltzer *et al.* 2005).

With the emergence of the concept of "ecological niche" (Grinnel, 1917), numerous studies focused on this aspect from different angles, contributing with theoretical arguments in order to facilitate understanding the matter. However, in 1958, Hutchinson's theory of hyper volume (1979, 1981), contributes with a new valuation of the concept of niche, which has allowed to fit its description and adequate quantification (Giller, 1984). Based on this theory species coexistence, as well as the subsequent interspecific relationships, are structured according to the level of segregation along the dimensions of the ecological niche.

Even though some researchers (Remmert, 1988) consider that the term niche should be deleted while some others think its connotation and scope should be indicated (Beltzer, 1981, 2003, Beltzer *et al.* 2005). Since its appearance to the present day, the interpretation of its signification has had different meanings. Though the concept is valid, it has determined discouragement due to the obstacles in some cases when trying to qualify its determining variables (Walker, 1987).

The partition of resources is closely related to the concept of niche. Due to the fact that this concept becomes inapplicable in practice in every possible dimension, it has been common that in the studies on the functional role played by the organismic units, three principal dimensions were selected (Pianka, 1973, 1975, 1982): food, space and time. This has given place to what Giller (1984) calls "differential overlapping of niches" by means of which when similar species overlap in one dimension, they are substantially separated in the other or others.

Within this framework, the knowledge of niches becomes really useful to become aware of the natural systems and to be able to anticipate qualitatively some impacts as a consequence of antrophic disturbances (Neiff, 1999).

As it was already discussed, the relationships established around the trophic resources play a key role within this conceptualization. In that sense, evidence obtained on the feeding of birds largely confirm the proposals presented by Root (1968) pointing out that the types of prey, their size and identification as well as capture efficiency are fundamental variables in the study of the trophic dimension of the niche and the interspecific relationships in a single guild (Orians, 1969; Cody, 1974; Wilson, 1974; Jordano, 1981; Nudd, 1983; Osborne *et al.*, 1983; Faaborg, 1985; Magurran, 1989; Kirkconnel *et al.*, 1992).

Guilds constitute one of the most important ecological issues in the study of a community. Although it is assumed that within a guild species use the resources in a similar way, this is not always the case (Landres and MacMahon, 1980) since, in addition to the kind of food, there are other components such as feeding site and preying habits (Wilson, 1974). In the same way, these aspects may change according to habitat and season. Another aspect to be considered is the size of the feeding group, i.e., if they tend to feed individually, in pairs or in unispecific or mixed flocks.

There is greater morphological and ecological similitude among the species of a single guild. Clearly, guilds are subgroups within communities, thus several guilds could be distinguished from the dietary point of view. The study of guilds is based on the analysis of the groups of species with similar ecological demands (Jaksic, 1981; Beltzer, 2003).

The principle of ecological incompatibility or competitive exclusion (Cabrera, 1932; Hutchinson, 1979, 1981) enunciates in its simplest manner that two or more species with similar ecological requirements (such as kind of food, nesting and sheltering sites) could not coexist indefinitely since sooner or later only the competitively superior species would survive. In that sense, in the measure that there are more individuals of one or both species in a given environment, the intensity of the competition would be increased. Consequently, competition within an assembly is more likely among similar species, particularly co-generic ones.

The mechanisms to avoid competition are a result of the joint evolution in shared habitats, but they also reveal pre-adaptations already established in their place of origin in the case of species of a more recent arrival in the American continent.

In general, ecological isolation by means of the partition of resources is observed, which is achieved through the capture of preys of different sizes and types, at different times of the day and in diverse microhabitats, using varied capture mechanisms. An additional factor which also allows coexistence of aquatic birds is the diversity and abundance of preys available along the year, activity hours and their escape behavior, which leads to a complex mosaic of possibilities. All of them are fundamental variables in the study of the trophic dimension of the niche and the trophic relationships among the species of a guild (Cody, 1974; Eckhardt, 1979; Jordano, 1981; Beltzer, 2003; Beltzer et al., 2005).

The aquatic ecosystems represent a good means to document such processes; however, the contributions made in the floodplain of the Parana River in that sense are scarce (Beltzer, 1995). In these systems, birds are outstanding consumers (Martinez, 1993) and among them, fishing birds possibly constitute the most diverse guild. In many cases –as for herons– the term "fishing" does not imply an exclusively ichthyophagous diet, since they can frequently capture a noticeable amount of insects, mollusks, crustaceans, arachnids, amphibians, reptiles, etc.

Among fishing birds, wading birds form a continuous series of species of similar morphology and capture techniques whose greater variation is expressed by body size. Nearly all of them capture their prey while wading or lurking in shallow water bodies. Thus, the feeding depth in each case is limited to the length of their legs and neck. In this group a relation between peak size and corporal weight is noticed so that species of a larger size can capture and intake bigger preys. In like manner, larger birds translocate at lower speed while they find their food, and they can even remain motionless for hours until the moment of capture. On the other hand, small and medium-sized herons are more agile and faster.

Literature is really vast regarding studies on trophic ecology of birds, not only concerning the association with habitats, feeding techniques and feeding spectrum (Block, et al., 1992; Landmann and Winding, 1995) but also in relation to herons in particular: Kushlan, 1976 a-b, 1978, 1981; Watmough, 1978; Whitfield and Cyrus, 1978; López-Ornat and Ramo, 1992; Tosi and Toso, 1979; Amat, 1984; Amat and Soriguer, 1981; Kushlan, 1981; Forbes, 1987; Rohwer, 1988; Amat and Aguilera, 1989; Frederick and Collopy, 1989; Wolf and Jones, 1989; Frederick et al., 1990; Marquis and Leitch, 1990; Erwin et al., 1991; Kersten et al., 1991; Shealer and Kress, 1991; Ewins and Hennessey, 1992; Lopez Ornat and Ramo, 1992; Moreira, 1992: Buckelew, 1993; Fernández Cruz and Campos, 1993; Gómez-Tejedor, 1993; Kelly, et al., 1993; Mc Neil, et al., 1993; OConnor, 1993; Ramo and Busto, 1993; Guillen, et *al.*, 1994; Maddock and Geering, 1994; Peris, *et al.* 1994 a-b; Lekuona and Campos, 1995; may be cited among others.

Summarizing, and bearing in mind the fact that birds constitute a characteristic feature of the aquatic environments throughout the world and that their specific composition is an indicator of the status of the water body, their productivity at the different trophic levels and the peculiarities of their structure and function; the study on ecological niches in their three dimensions is approached: a) temporal sub niche, b) spatial sub niche and c) trophic sub niche of the species of the Ardeidae family, their interspecific relationships and isolation mechanisms which allow their coexistence and avoid competitive exclusion. This aspect becomes particularly relevant since in the floodplains of great rivers, the spatial and temporal variability patterns are strongly conditioned by the hydrosedimentological regime (Junk *et al.*, 1989; Neiff, 1990). The direct hydrosedimentological influence of the river on the structure of the vegetation (changes in the specific richness, biomass and replacement of plant bioforms) presents direct changes in the offer of habitats for birds (Beltzer y Neiff, 1992; Higgins *et al.*, 1996).

On the basis of all that was expressed, the need to clarify the following issues arises:

- Is there a similar pattern in the feeding techniques among the heron species considered?
- Do isolation mechanisms vary in relation to the composition of the trophic spectrum?
- Is the effective niche expressed on an annual or seasonal basis as another expression of the isolation mechanisms?
- Do the spatial (spatial sub niche) and temporal (temporal sub niche) dimensions have a similar value to the trophic one as a mechanism of segregation?
- Is there agreement between the feeding ecology and the morphological design of the herons considered in the present study?

AREA OF STUDY AND METHODS

The great Paraná basin (2,600,000 Km^2) covers a large portion of the Neotropical Realm, extending from the Andes to the Atlantic coast and connecting the central regions of South America along 2,200 Km. in a north–south direction. Most of the basin is under subtropical climates, which range from desertic in the west to humid in the east (Iriondo *et al.* 2007)

The Middle Paraná extends 600 Km. in the heart of the interior lowlands of the continent. Its flood plain is large, complex and develops on the right bank and extends from near the confluence of the Paraná and Paraguay rivers to the city of Diamante (Entre Ríos). Corresponds to the Chaqueña subregion occupying the northern and central Argentina, southern Bolivia, western Paraguay and central and northwestern Brazil (Morrone, 2001) and involves several ecosystems. The essential feature of these plains is to be periodically flooded by the increasing annual of Paraná River.

Fieldwork was conducted on the Carabajal island, Santa Fe, northern Argentina (31°39'S, 60°42'W) that has an area of about 4,000 ha. (Fig. 1). This island belongs to the geomorphologic unit called "plain of banks" (Iriondo and Drago, 1972). Numerous lenitic water bodies (stagnant waters) are found in this island, some of considerable extension such as ponds "La Cuarentena" (80 ha.), "La Cacerola" (80 ha.), "Vuelta de Irigoyen" (70 ha.) and

"El Puesto" (40 ha.). For this study, and following the criteria proposed by Beltzer (1981, 1990a, 1990b, 1991), Neiff (1975, 1979, 1986a, 1986b) and Beltzer and Neiff (1992) for the flood valley of the Paraná river, the following units of vegetation and environment ("UVEs") have been recognized: open waters, floating and rooted aquatic vegetation, gallery forests, grasslands, pastures, beach and forest.

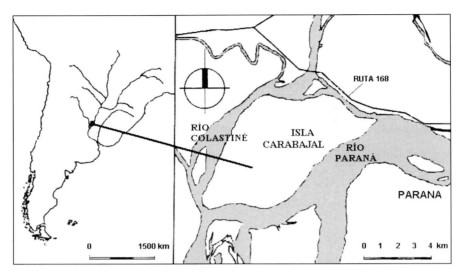

Figure 1. Geographical location of the Carabajal island, Province of Santa Fe, Argentina. (grey = water, white = land).

The species of Ardeidae considered for this study were:

Scientific name	Common name	Samples
Ardea cocoi (Linnaeus, 1766)	Cocoi Heron	29
Butorides striatus striatus (Vieillot, 1817)	Striated Heron	51
Bubulcus ibis ibis (Linnaeus, 1758)	Cattle Egret	30
Tigrisoma lineatum marmoratum (Vieillot, 1817)	Rufescent Tiger-Heron	17
Syrigma sibilatrix sibilatrix (Temmminck, 1824)	Whistling Heron	19
Egretta thula thula (Molina, 1782)	Snowy Egret	31
Ixobriyhus involucris (Vieillot, 1823)	Stripe-Backed Bittern	13
Ardea alba (Gmelin, 1798)	Great Egret	38
Nycticorax nycticorax hoactli (Gmelin, 1789)	Black-Crowned Night-Heron	40

The individuals captured with a firearm and some few caught with mist nets between 1999 and 2002 were used. Stomach contents from alive birds were obtained by stomach washing following Emison (1968) and Cowan (1983). It operates by forcing water into the proventriculus through a plastic tube, after which birds were inverted, pressure applied to the stomach and aimed the bird at a container. Stomachs from corpses were injected with 10% formalin (to stop digestive processes) while in the field and opened in the laboratory. All contents were fixed in 10% formalin for subsequent qualitative and quantitative analysis. The hour of capture and the weight of the birds and their stomachs were recorded. Field observations were also conducted to determine the habitats used and the hours of activity.

Once in the laboratory, stomach contents were analyzed individually under a binocular magnifying glass, to identify and quantify the organisms at different levels of taxonomic resolution. In order to count organisms in advanced digestion state, the key structures or pieces such as heads, jaws, elytra, chelicerae, etc, were regarded as individuals.

The contribution of each prey item to the diet of the species was established by applying the index of relative importance (IRI; Pinkas *et al.*, 1971):

$$IRI = \%FO \cdot (\%N + \%V)$$ (1)

Where: %FO is the percentage frequency of occurrence of a particular category of food, %N is the percentage numerical and %V the percentage by volume (measured by water column displacement when all items from a single food category are introduced into a test tube).

Trophic diversity was determined following Hurtubia's criterion (1973) to calculate the diversity (H) of prey for each individual using the formula of Brillouin (1965):

$$H = (\frac{1}{N}).(\log_2 N! - \sum \log_2 N_i!)$$ (1)

Where: N is the total number of organisms found in the stomach of each individual and Ni is the total number of preys in each stomach. The accumulated trophic diversity (Hk) was obtained by randomly adding trophic diversity's values (H) per stomach. The asymptote (point t, p.t.) of the curve, that results of its graphic representation, allows to determine the minimum sample size.

Similarities and degrees of diet overlapping were estimated by using a percentage of similarity according to Colwel y Futuyma (1971):

$$Cih = 1 - \frac{1}{2} \sum j \, /pij - phi/$$

Where: Cih is niche's proportional overlapping species and Pj represents the importance of j species in i and h specie's diet respectively.

The trophic amplitude of the niche was calculated by means of the index of Levins (1968):

$$Nb = (\sum p_{ij}^2)^{-1}$$ (5)

where: P_{ij} is the probability of item i in the sample j. It was calculated for each season to analyze the seasonal equivalent of the diet.

Feeding efficiency, Pe, was estimated following Acosta Cruz *et al.* (1988) and calculated per each year's season:

$$Pe = 1 \frac{x \, p. \, cont. \, x \, 100}{x \, p. \, corp.}$$ (4)

where: p. cont. is the weight of the stomach contents (in g) and p. corp. is the weight of the body of birds (in g).

Dietary selectivity was evaluated applying the Spearman Rank Correlation, rs (Sokal and Rohlf, 1979; Schefler, 1981):

$$r_s = 1 - \left[6\sum (X-Y)^2 \Big/ n(n^2 - 1) \right]$$

(3)

where: X is the abundance range of prey found in the stomach, Y is the abundance ordinal range of the prey in the study area, according a qualitative evaluation and n is the number of species prey.

With the purpose of establishing the hourly rhythm of the feeding activity, the average satiety index was calculated, IF (Mean Index of Fullnes, Maule and Horton, 1984):

$$IF = \left[\overline{x}\ cont.vol\ (ml) \Big/ \overline{x}\ corp.weight\ (g) \right].100$$

(6)

where: Cont. vol. is the volume of the stomach contents (in cm3) and corp. weight m is the body mass of the bird to each time interval of capture (in g).

Finally, the association of this species with different environments typical of the flood valley of the Paraná River was analyzed by means of the index of habitat preference, Pi (Duncan, 1983).

$$Pi = \log \cdot \left[Vi \Big/ Ai \right] + 1$$

(7)

Where: Vi is the percentage of individual recorded in each "UVEs" and Ai is the percentage of cover corresponding at each "UVEs". Following the criteria proposed by Bignal *et al.* (1988) values higher of 0.3 indicate high preference for one specific "UVEs" and values lower indicate a smaller preference.

Trophic relationships were compared by using an UPGMA grouping analysis where Sorensen´s similarity index was applied. Each species represented an operative taxonomic unit (OTU). Analysis was performed with MVSP software (Kovach, 1999). Coefficient's value varies between 0 and 1, being 1 the value assigned to a maximum similarity.

RESULTS

Trophic Spectrum

Cocoi heron

All studied stomachs contained food. The trophic spectrum was composed by 17 taxonomic entireties, been fishes the most frequent food item, followed by reptiles, amphibians and mammals.

Ardea cocoi's trohpic spectrum. N= amount of individuals; F= capture frequency; H= habitat; n.i.= non identified items; A= aquatic, T= terrestrial, ? = unknown.

	N	F	H
Mammalia			
Rodentia			
Holochilus brasiliensis	1	1	A
Amphibia			
Leptodactyllidae			
Leptodactillus ocellatus	3	2	A
Hylidae			
Hyla pulchella	5	3	A
Pisces			
Characidae			
Astyanax bimaculatus	6	9	A
Salminus maxillosus	7	4	A
Serrasalmidae			
Serrasalmus spilopleura	2	1	A
Erythrinidae			
Hoplias malabaricus	26	25	A
Anostomidae			
Leporinus obtusidens	23	17	A
Curimatidae			
Prochilodus lineatus	32	30	A
Pimelodidae			
Pimelodus albicans	13	12	A
Callichthyidae			
Hoplosternum littorale	11	9	A
Loricariidae			
Loricaria anus	1	1	A
Synbranchidae			
Synbranchus marmoratus	1	1	A
n.i.	19	20	A
Insecta			
Hemiptera			
Belostomidae			
Belostoma sp.	9	5	A
Orthoptera			
Paulinidae			
Paulinia acuminata	3	2	A
Crustaceoa			
Decapoda			
Trichodactyllidae			
Trichodactyllus borelianus	6	4	A

Striated heron

All stomachs contained food. The spectrum was composed of fifty one (51) taxonomic entities. Fishes and insects were the better represented items while crustaceans, arachnids and amphibians were less common.

Butorides striatus striatus' trohpic spectrum. N= amount of individuals; F= capture frequency; H= habitat; n.i.= non identified items; A= aquatic, T= terrestrial, ? = unknown.

N	F	H		
Amphibia				
Leptodactylidae				
Leptodactyllus ocellatus	2	2	A	
Hyliidae				
Hyla pulchella	10	12	A	
Pisces				
Characiformes				
Characidae				
Holohestes pequira	97	45	A	
Prionobrama paraguayensis	5	2	A	
Astyanax bimaculatus	8	5	A	
Astaynax sp.	3	2	A	
Salminus maxillosus	6	4	A	
n.i.	5		4	A
Serrasalmidae				
Serrasalmus spilopleura	1	1	A	
Parodontidae				
Apareiodon affinis	3	2	A	
Curimatidae				
Prochilodus lineatus	6	4	A	
Pseudocurimata sp.	1	1	A	
Erythrinidae				
Hoplias malabaricus	3	2	A	
Rhamphichthyidae				
Eigennmania virescens	6	4	A	
Rhamphicthys rostratus	1	1	A	
Pimelodidae n.i.	1	1	A	
Callichthyidae				
Hoplosternun thoracatum	3	2	A	
H. littorale	5	2	A	
Cyprinodontidae				
Pterolebias longipinnis	2	2	A	
Cichlidae				
Crenicichla sp.	1	1	A	
Aequidens portalegrensis	6	4	A	
Aranae				
Pisauridae n.i.	3	2	A	
Lycosidae n.i.	3	2	A	
n.i.	17	12	?	
Crustacea				

(Continued)

Amphypoda				
Hyalellidae				
Hyalella curvispina	5	3	A	
Decapoda				
Trichodactyilidae				
Trichodactyllus borelianus	2	2	A	
Palaemonidae				
Macrobrachium borelli	10	7	A	
Palaemonetes argentinus	21	10	A	
Insecta				
Odonata n.i.	30	15	T	
Orthoptera				
Acrididae	9	5	T	
n.i.	12	5	T	
Hemiptera				
Notonectidae n.i.	2	1	A	
Belostomidae				
Belostoma micantulum	8	5	A	
B. discretum	1	1	A	
Belostoma sp.	2	2	A	
Corixidae				
Tricocorixa sp.	6	4	A	
n.i.	1	1	A	
Coleoptera				
Carabidae n.i.	3	2	T	
Curculionidae n.i.		5	3	A
Dytiscidae n.i.	3	2	A	
Hydrophilidae n.i.	9	8	A	
Ephemeroptera				
Baetidae				
Baetis sp.	1	1	A	
Diptera				
Chironomidae n.i.	12	1	A	
Muscidae n.i.	5	1	T	
Hymenoptera				
Ponerinae	1	1	T	
n.i.	15	7	T	

Cattle egret

All stomachs contained food. The spectrum was composed by seventeen (17) taxonomic entities, mostly represented by insects. The Orthoptera (Family Pulinidae: *Marellia* sp., *Paulinia acuminate* and Leptismidae: *Cornops acquaticum*) were the better represented items, followed by Coleoptera (Hydrophilidae, Dytiscidae and Curculionidae). Arachnida y Amphibia were the remaining items.

Bubulcus ibis ibis' trohpic spectrum. N= amount of individuals; F= capture frequency; H= habitat; n.i.= non identified items; A= aquatic, T= terrestrial, ? = unknown.

N	F	H	
Organismos			
INSECTA			
Orthoptera			
Paulinidae			
Marellia sp.	35	42	A
Paulinia acuminata	13	23	A
Lepismidae			
Cornops aquaticum	85	54	A
Acrididadae (n.i.)	29	15	T
Orthoptera (n. i.)	12	21	?
Gryllidae			
Gryllodes sp.	31	12	T
Coleoptera			
Hydrophilidae n.i.	10	8	A
Trotispernus sp.	4	2	A
Dytiscidae n.i.	12	9	A
Curculionidae	13	3	A
n.i.	2	1	?
Hemiptera			
Belostomidae			
Belostoma sp.	3	1	A
Lepidoptera (larvas n.i.)	1	1	?
Aranae			
Pysauridae n.i.	41	13	A
Amphibia			
Hyllidae			
Hyla pulchella	12	9	A
Semillas n.i. sp. A	5	3	?
Semillas n.i. sp. B	1	1	?

Rufescent tiger-heron

All stomachs contained food (n= 17) where fishes were the most abundant item (including species like *Synbranchus marmoratus* and *Hoplosternun littorale*), followed by insects, amphibians and reptiles.

Tigrisoma lineatum marmiratum's trohpic spectrum. N= amount of individuals; F= capture frequency; H= habitat; n.i.= non identified items; A= aquatic, T= terrestrial, ? = unknown.

N	F	H	
Reptilia			
Colubriformes			
Colubridae n.i.	2	2	A
Amphibia			
Leptodactyllidae			
Leptodactillus ocellatus	2	1	A
Hyliidae			
Hyla pulchella	4	2	A
Pisces			

Siluriformes			
Callichthyidae			
Hoplosternum littorale	12	9	A
Synbranchiformes			
Synbranchidae			
Synbranchus marmoratus	23	13	A
Caraciformes			
Caracidae			
Astyanax sp.	12	9	A
Holoshestes pequira	28	14	A
Insecta			
Odonata			
Anisoptera	4	2	T
Zygoptera	1	1	T
Coleoptera			
Hydrophilidae			
Tropisternus sp.	2	2	A
Hemiptera			
Belostomidae			
Belostoma sp.	3	2	A

Whistling heron

All stomachs contained food. The spectrum was composed by fourteen (14) taxonomic entities, mostly represented by insects (forms associated to plants and terrestrial environments). The remaining groups (Colubridae, Hylidae and Crustacea) were less represented.

Syrigma sibilatrix sibilatrix's trohpic spectrum. N= amount of individuals; F= capture frequency; H= habitat; n.i.= non identified items; A= aquatic, T= terrestrial, ? = unknown.

N	F	H	
Reptilia			
Colubridae n.i.	5	3	T
n.i.	2	1	T
Amphibia			
Hylidae			
Hyla pulchella	6	4	A
Pisces			
n.i.	2	1	A
Aranae			
Lycosidae n.i.	9	5	T
n.i.	12	8	T
Crustacea			
Decapoda			
Trichodactylidae			
Trichodactyllus borelianus	5	3	A
Insecta			
Odonata			
Anisoptera n.i.	15	9	T
Orthoptera			

(Continued)			
Acrididae			
Cornops aquaticum	23	11	A
Hemiptera			
Belostomatidae n.i.	7	5	A
Coleoptera			
Dytiscidae n.i.	11	8	A
Carabidae			
Diloboderus abderus	2	1	T
n.i.	21	12	T
n.i.	34	16	?

Snowy egret

All stomachs contained food. The spectrum was composed by thirty (30) taxonomic entities, been fishes the most remarkable item.

Egretta thula thula's trohpic spectrum. N= amount of individuals; F= capture frequency; H= habitat; n.i.= non identified items; A= aquatic, T= terrestrial, ? = unknown.

N	F	H	
Pisces			
Characidae			
Aphyocharax rubropinnis	15	2	A
Astyanax bimaculatus	3	2	A
Astyanax sp.	3	2	A
Holoshestes pequira	21	27	A
Hyphessobrycon anisitsi	30	28	A
Odontostilbe paraguayensis	13	7	A
Serrasalmidae			
Serrasalmus spilopleura	3	2	A
Erythrinidae			
Hoplias malabaricus	2	2	A
Anostomidae			
Schizodon fasciatum	6	4	A
S. platae	2	1	A
Curimatidae			
Prochilodus lineatus	1	1	A
Curimatorbis platanus	5	3	A
Rhamphichtyidae			
Eigenmania virescens	12	8	A
Pimelodidae			
Pimelodus albicans	1	1	A
Callichthyidae			
Corydoras hastatus	9	7	A
Poeciliidae			
Cnesterodon decenmaculatus	3	3	A
n.i.	2	1	A

Trophic Relationships and Mechanisms of Ecological Segregation...

Synbranchidae			
Synbranchus marmoratus	3	2	A
Cichlidae			
Apistogramma combrae	5	3	A
n.i.	2	1	A
Lebiasinidae			
Pyrrhulina australis	1	1	A
Insecta			
Hemiptera			
Belostomidae			
Belostoma micantulum	3	2	A
Coleoptera			
Hydrophilidae			
Tropisternus sp.	6	4	A
Dytiscidae n.i.	1	1	A
Odonata n.i.	4	4	T
Diptera n.i.	25	1	T
Aranae n.i.	2	1	T
Crustacea			
Palaemonidae			
Palaemonetes argentinus	1	1	A
Machrobrachium borelli	3	2	A
Hyallelidae			
Hyalella curvispina	2	1	A

Stripe-backed bittern

All stomachs contained food. The spectrum was composed by fourteen (14) taxonomic entities where insects like Odonata, Coleoptera, Hemiptera and Orthoptera were the most frequent. Fishes and Arachnids showed lower values.

Ixobrychus involucris's trohpic spectrum. N= amount of individuals; F= capture frequency; H= habitat; n.i.= non identified items; A= aquatic, T= terrestrial, ? = unknown.

N	F	H		
Pisces				
n.i.	2	1	A	
Aranae				
Pisauridae n.i.	1	1	T	
n.i.	3	2	?	
Insecta				
Orthoptera				
Paulinidae				
Marellia sp.	8	5	A	
Paulinia acuminata	5	3	A	
Lestismidae				
Cornops aquaticum	6	6	A	

(Continued)				
Odonata				
Anisoptera n.i.	6	3	T	
Zigoptera n.i.	10	6	T	
Coleoptera				
Cucurlionidae n.i.		6	5	A
Dytiscidae n.i.	7	3	A	
Hydrophilidae n.i.	11	7	A	
Hemiptera				
Belostomatidae				
Belostoma ap.	6	5	A	
n.i.	10	8	A	
Insecta n.i.	22	10	?	

Great egret

All stomachs contained food. The spectrum was composed of thirty three (33) taxonomic entities. Fishes and insects were better represented.

Casmerodius albus' trohpic spectrum. N= amount of individuals; F= capture frequency; H= habitat; n.i.= non identified items; A= aquatic, T= terrestrial, ? = unknown.

N	F	H		
Pisces				
Characidae				
Aphiocarax rubripinnis	8	5	A	
Astyanax fasciatus	3	2	A	
Astyanax sp.	13	9	A	
Salminus maxillosus	4	2	A	
Serradalmidae				
Serrasalmus spilopleura	1	1	A	
Anostomidae				
Leporinus obtusidens	10	7	A	
Schizondon fasciatum	4	2	A	
Hemiodidae				
Appareiodon affinis	3	2	A	
Characidium fasciatum	1	1	A	
Curimatidae				
Prochilodus lineatus	6	3	A	
Curimatorbis platanus	29	18	A	
Gasterostomus latior	3	1	A	
n.i.	32	6	A	
Rhamphichtyidae				
Hypopomus brevirostris	1	1	A	
Pimelodidae				
Pimelodus albicans	5	3	A	

n.i.	3	1	A	
Callichthyidae				
Corydoras hastatus	2	1	A	
Corydoras sp.	1	1	A	
Hoplosternum thoracatum		7	4	A
Cyprinodontidae				
Pterolebias longipinnis	1	1	A	
Cichlidae				
Apistogramma combrae	1	1	A	
n.i.	3	3	A	
Lesbiasinidae				
Pyrrhulina australis	13	7	A	
Insecta				
Hemiptera				
Belostomidae				
Belostoma micantulum	5	3	A	
Belostoma sp. 1	1	1	A	
Orthoptera				
Acrididae n.i.	1	1	T	
Paulinidae				
Paulinia acuminata	3	2	A	
Coleoptera				
n.i.	3	3	A	
Hymenoptera	1	1	A	
Aranae n.i.	2	1	T	
Crustacea				
Decapoda				
Trichodactyllidae				
Trichidactyllus borelianus	1	1	A	
Palaeminidae				
Machrobrachium borelli	5	3	A	
Palaemonetes argentinus	1	1	A	

Black-crowned night-heron

All stomachs contained food. The spectrum was composed by sixteen (16) taxonomic entities. Fishes, especially *Synbranchus marmoratus*, were the most abundant items, followed by Hylidae, Insecta and Crustacea.

Nycticorax nycticorax hoactli's trohpic spectrum. N= amount of individuals; F= capture frequency; H= habitat; n.i.= non identified items; A= aquatic, T= terrestrial, ? = unknown.

N	F	A	
Amphibia			
Hylidae			
Hyla pulchella	12	9	A
Pisces			
Characidae			
Astyanax bimaculatus	3	2	A
Astyanax sp.	4	2	A
Anostomidae			
Leporinus obtusidens	4	3	A
Erythrinidae			
Hoplias malabaricus	3	2	A
Curimatidae			
Pseudocurimata sp.	2	2	A
Ramphichthyidae			
Eigenmannia virescens	1	1	A
Pimelodidae			
Pimelodella maculatus	1	1	A
Pimelodus albicans	8	5	A
Callichthyidae			
Hoplosternun littorale	7	6	A
Synbranchidae			
Synbranchus marmoratus	21	19	A
Loricariidae			
Hypostomus sp.	3	3	A
Insecta			
Hemiptera			
Belostomidae			
Belostoma micantulum	8	5	A
Orthoptera			
Acrididae n.i.	2	1	T
Crustacea			
Trychodacthillidae			
Trychodactyllus borelianus	10	6	A
Palaemonidae			
Palaemonetes argentinus	9	5	A

RELATIVE IMPORTANCE INDEX (IRI)

Cocoi Heron

Fishes were diet's most important component in quantity, volume and frequency of occurrence. More abundant preys (including some species of economic interest) were *Hoplias malabaricus*, *Leporinus obtusidens* and *Prochilodus lineatus*. These items represented the most basic food category. A secondary category of food was composed by other vertebrate preys like *Holochilus brasiliensis* (just one individual found) and amphibians (*Leptodactyllus ocellatus* and *Hyla pulchella)*. Insects were poorly represented, then being considered as an accessory category of food.

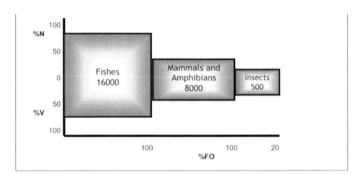

Ardea cocoi. %N = percentage numerical, %V = percentage by volume, % FO = percentage frequency of occurrence of a particular category of food.

Striated heron

Trophic spectrum's higher percentage was represented by fishes, constituting the basic food category. *Holoshestes pequira* was the most abundant fish species, followed by *Prionobrama paraguayensis, Astyanax bimaculatus, Astyanax sp., Salminus maxillosus, Prochilodus lineatus, Hoplosternum littorale* and *Aequidens portalegrensis*. The remaining species (*Apareiodon affinis, Hoplias malabaricus, Eigennmania virescens, Pterolebias logipinnis*) were poorly represented. Insects belonged to aquatic and terrestrial forms, being Belostomidae, Corixidae and Chironomidae the most abundant within the former and Odonata and Orthoptera within the last.

At the same time, seeds and rests of macrophites were found but considered as non voluntary ingested items, since they may be eaten when trying to capture animal preys.

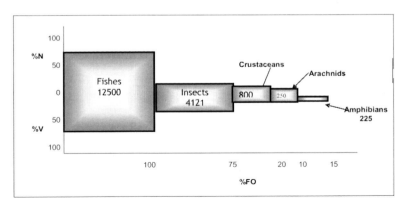

Butorides striatus. %N = percentage numerical, %V = percentage by volume, % FO = percentage frequency of occurrence of a particular category of food.

Cattle egret

Insects constituted the main food category, basically represented by the Orthoptera with vegetation's associated forms like *Marelia* sp., *Paulinia acuminata* and *Cornops aquiaticum*. Among Coleopterans, Hydrophilidae, Dytiscidae and Curculionidae must be highlighted. The Aranae and Amphibians constituted secondary food categories, respectively represented by Pysauridae and *Hyla pulchella*.

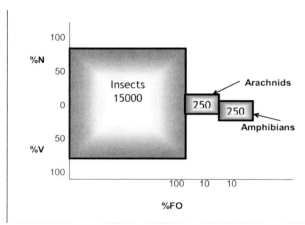

Bubulcus ibis. %N = percentage numerical, %V = percentage by volume, % FO = percentage frequency of occurrence of a particular category of food.

Rufescent tiger-heron

Fishes like *Holoshestes pequira*, *Synbranchus marmoratus* and *Hoplosternum littorale* represented the basic diet. Reptilia and Amphibia were considered as a secondary food category and were represented by Colubridae (not indentified), *Leptodactyllus ocellatus* and *Hyla pulchella*. An accessory food category was composed by insects (i.e. Odonata, Coleoptera y Hemiptera).

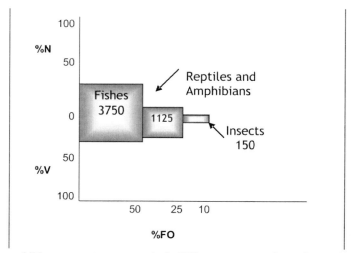

Tigrisoma lineatum. %N = percentage numerical, %V = percentage by volume, % FO = percentage frequency of occurrence of a particular category of food.

Whistling heron

Insects represented the main food resource, out of which, aquatic vegetation associated forms of Orthopterans like *Cornops aquaticum* should be noted. Besides, some aquatic and terrestrial Coleoptera as well as the Aranae, Crustacea and Vertebrata were considered as secondary food categories.

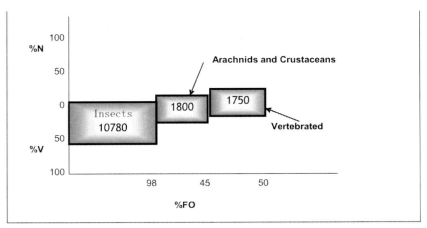

Syrigma sibilatrix. %N = percentage numerical, %V = percentage by volume, % FO = percentage frequency of occurrence of a particular category of food.

Snowy egret

Fishes were the most important items in volume and quantity, especially those from family Characidae. Fishes found were mostly forms associated to aquatic vegetation. Species like *Astyanax bimaculatus* and *Curimatorbis platanus* can be found in open waters, while others (*Holoshestes pequira*, *Hyphesobrychon anisitsi* and *Odontostilbe paraguayensis*), can be also captured close to aquatic vegetation (Cordiviola de Yuan, 1980). Insects constituted a secondary food category and Crustaceans were accessory items.

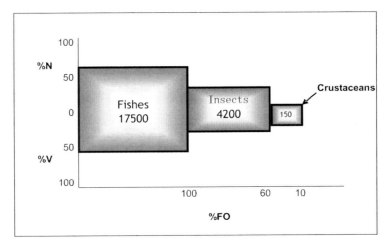

Egretta thula. %N = percentage numerical, %V = percentage by volume, % FO = percentage frequency of occurrence of a particular category of food.

Stripe-backed bittern

Insects represented the basic diet, especially those associated to aquatic vegetation like *Paulinia acuminata*, *Marelia* sp. and *Cornops aquiaticum*. Coleopterans like Curculionidae, Dytiscidae and Hydrophilidae were followed in importance by the Belostomatidae. The Odonata were the only terrestrial forms of the spectra. Pisces and Arachnida were poorly represented, considered then as an accessory food item.

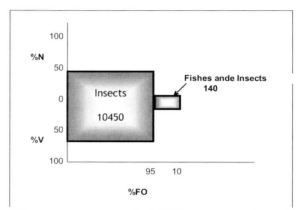

Ixobrychus involucris. %N = percentage numerical, %V = percentage by volume, % FO = percentage frequency of occurrence of a particular category of food.

Great egret

Fishes constituted the basic diet, being highly represented (in number and volume) by *Prochilodus lineatus*, *Pimelodus albicans*, *Leporinus obtusidens*, *Serrasalmus spilopleura* y *Astyanax sp*. The Curimatidae were the most frequent and abundant species. Most fish's species like *Hipopomus brevirrostris* y *Apistogramma combrae* are typical of areas covered with aquatic vegetation. Others are usually found in open waters like the Curimatidae and *Apereiodon affinis* (Beltzer y Oliveros, 1982). Within insects, some aerial and terrestrial forms like Odonata (adults), Hymenoptera and Orthoptera were found, constituting a secondary food category.

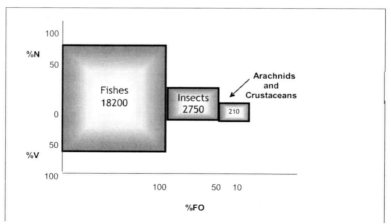

Casmerodius albus. %N = percentage numerical, %V = percentage by volume, % FO = percentage frequency of occurrence of a particular category of food.

The low volume presented by the animal fraction had, on the other hand, a high percentage of occurrence. Still it was considered as an accidental ingestion. *Riciocarpus natans* was one of the best represented species, especially in those cases where it was abundant at the feeding areas where the bird was captured.

Black-crowned night-heron

Fishes were predominant food items, represented by juveniles of *Leporinus obtusidens, Pimelodus clarias, Hoplias malabaricus, Hypostomus* sp. and adults of *Synbranchus marmoratus* (being this species the most abundant and frequent).

Within insects Belostomatidae and Acrididae were frequently found, while the same happened with Trychodactilidae and Palaemonidae within crusteaceans.

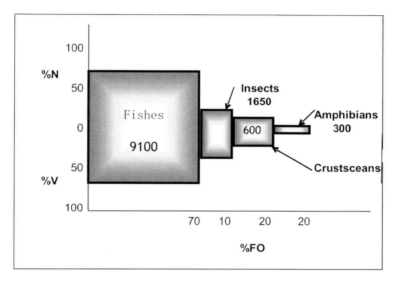

Nycticorax nycticorax. %N = percentage numerical, %V = percentage by volume, % FO = percentage frequency of occurrence of a particular category of food.

TROPHIC DIVERSITY

Cocoi Heron

Diversity values ranged from 0.45 to 3.23, being the stomachs comprised in the mean diversity interval the most frequent. Accumulated trophic diversity (Hk) was 1.78, showing the asymptote that allows pointing out that the minimum sample was reached (Magurran, 1989).

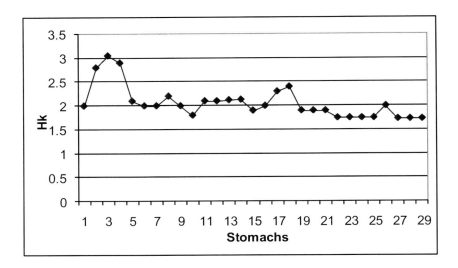

Striated heron

Trophic diversity values ranged from 0 to 2.99, being the stomachs comprised in the mean diversity interval the most frequent. Accumulated trophic diversity (Hk) was 3.89, showing the asymptote that allows pointing out that the minimum sample was reached (Magurran, 1989).

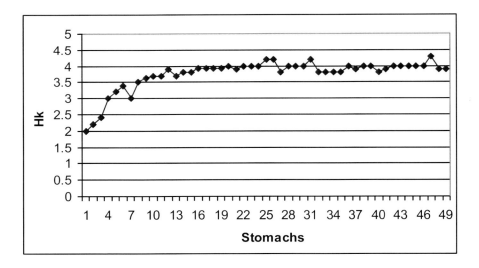

Cattle egret

Diversity values per stomachs ranged from 1.03 to 2.76, being the comprised in the mean diversity interval the most frequent. Accumulated trophic diversity (Hk) was 2.77, getting the pt (Magurran, 1989), which allows pointing out that it has worked with the minimum adequate sample.

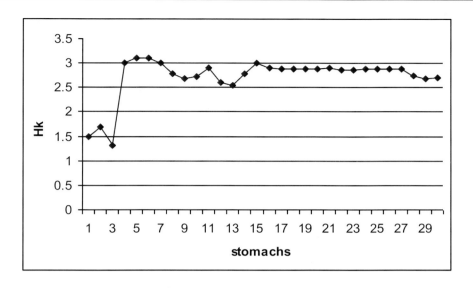

Rufescent tiger-heron

Diversity values ranged from 0.12 to 2.43, being the stomachs comprised in the mean diversity interval the most frequent. Accumulated trophic diversity (Hk) reached a value of 2.15

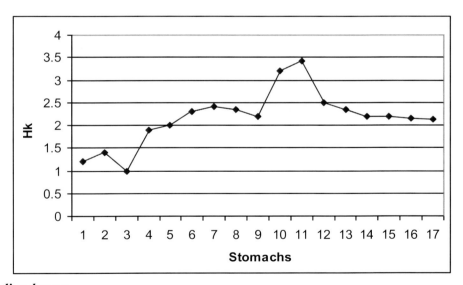

Whistling heron

Diversity values ranged from 0.23 to 2.15, being the stomachs comprised in the mean diversity interval the most frequent. Accumulated trophic diversity (Hk) was 1.70. With twenty samples, the curve asintotiza (pt), which has worked with a minimum sample properly.

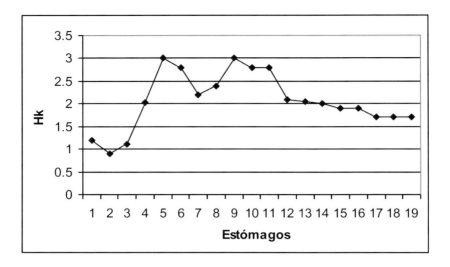

Snowy egret

Diversity values per stomachs ranged from 1.1. to 2.78, being the most frequent those comprised in the low diversity interval. Accumulated trophic diversity (Hk) was 1.20

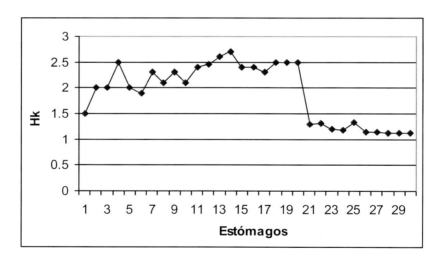

Stripe-backed bittern

Diversity values ranged from 0.22 to 2.89, being the stomachs comprised in the mean diversity interval the most frequent. Accumulated trophic diversity (Hk) reached a value of 1.58 obtained by applying Magurran's criterion (1989), despite the low number of stomachs examined for this species

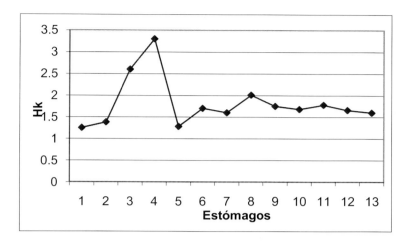

Great egret

Trophic diversity values per stomach ranged from 0.21 to 2.34, being the most frequent those comprised in the mean diversity interval, while accumulated trophic diversity (Hk) was 1.86 (p. t, Magurran, 1989).

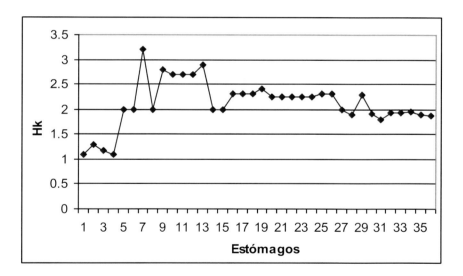

Black-crowned night-heron

Diversity values ranged from 0 to 1.86, being the stomachs comprised in the low diversity interval the most frequent. Accumulated trophic diversity (Hk) reached a value of 3.5 obtained by applying Magurran's criterion (1989), which allows pointing out that a minimum quantitative sample was used.

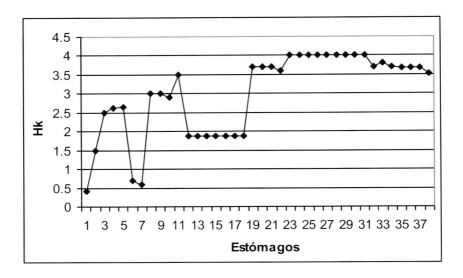

DIET OVERLAP

The comparison of diets in pairs allowed for the determination of similarities and degrees of overlapping. The results obtained were the following:

	B	C	D	E	F	G	H	I
A	12	3	2	25	13	2	4	0,8
B	--	2	5	19	24	4	8	2
C	--	--	0,8	6	5	18	3	7,5
D	--	--	--	10	12,5	3	6,7	2,3
E	--	--	--	--	16,4	6,1	5,8	1,8
F	--	--	--	--	--	3,9	7,2	0,9
G	--	--	--	--	--	--	0,82	1,3
H	--	--	--	--	--	--	--	0,31

A= Cocoi Heron; B= Striated Heron; C= Cattle Egret; D= Rufescent Tiger-Heron; E= Great Egret; F= Snowy Egret; G= Whistling Heron; H= Black-Crowned Night-Heron; I= Stripe-Backed Bittern.

The most important trophic relationship corresponded to the pairs of big and medium sized herons and basically those ictiophagous herons such as Cocoi Heron / Great Egret and Snowy Egret/Striated Heron, showing values ranging from 25, 24 to 19 % overlap. Likewise, values were relatively high in the case of herons frequently found in grassland, such as Cattle Egret/ Whistling Heron.

On the other hand, the remaining pairs and those showing low values are expressed among the pairs of ardeids closely related to open water environments, aquatic vegetation both floating as well as rooted such as Striated Heron/ Cattle Egret; Cocoi Heron/ Whistling Heron. Stripe-Backed Bittern with various species like Black-Crowned Night-Heron, Whistling Heron, Snowy Egret with the lowest overlapping values.

Trophic Relationships and Mechanisms of Ecological Segregation…

TROPHIC NICHE AMPLITUDE

As for the seasonal equivalent of the diet of herons measured through the trophic niche amplitude the values were as follows:

Specie	Spring	Summer	Autumn	Winter
Cocoi Heron	3.87	4.21	2.89	1,98
Striated Heron	2.56	3.45	-	-
Cattle Egret	3.45	2.09	1.98	2.01
Rufescent Tiger-Heron	2.33	3.2	2.25	3.42
Whistling Heron	1.98	2.34	2.21	1.99
Snowy Egret	2.83	3.03	7.39	4.4
Stripe-Backed Bittern	2.34	3.53	3.21	2.87
Great Egret	3.86	5.57	2.94	3.95
Black-Crowned Night-Heron	5.22	3.84	2.55	3.47

Note: The lack of values for autumn and winter for the Striated Heron is due to its status as a migratory species, whose presence in the area shown in the spring and summer.

DIETARY EFFICIENCY

Regarding dietary efficiency the recorded values were the following:

Specie	Spring	Summer	Autumn	Winter
Cocoi Heron	99.9	98.87	89.9	91.02
Striated Heron	97.56	96.89	--	--
Cattle Egret	89.05	93.56	94.03	91.9
Rufescent Tiger-Heron	90.0	89.9	91.34	94.56
Whistling Heron	96.01	98.3	89.9	92.34
Snowy Egret	98.03	99.01	98.2	99.3
Stripe-Backed Bittern	88.9	91.45	91.04	88.92
Great Egret	99.31	99.11	98.80	99.0
Black-Crowned Night-Heron	99.00	97.32	92.3	97.82

SIZE OF PREY

Cocoi Heron

The size of the prey consumed varied between 44 and 280 mm. The range 151 -> 200 mm. was the most represented, which include both adults and juveniles fish of *Hoplias malabaricus*, *Leporinus obtusidens*, *Prochilodus lineatus* and mammals as *Holochilus brasiliensis*. The smaller preys were insects and the intermediate size range was occupying by fish (*Astyanax* sp.) and some insects.

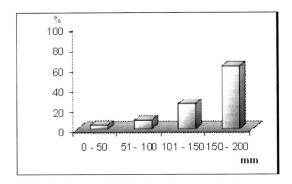

Striated heron

The prey consumed by this species varied in a small range (8 and 50 mm.). Those whose sizes correspond to the interval 21-30 mm. were most abundant (accounting for 60%), among which include fish, such as *Holoshestes pequira, Astyanax bimaculatus, Apareiodon affinis* and juvenile of *Prochilodus lineatus, Hoplias malabaricus*. Bigger preys (7%) were represented by the amphibians with *L. ocellatus*, while those between 10 and 20 mm. were mostly insects and arachnids.

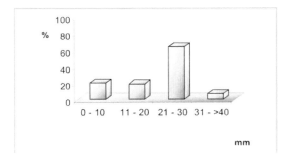

Cattle egret

Prey ingested varied in size with the 21-30mm. size class being the most numerous, representing 45% of the whole diet (Fig. 3). Different species of Orthoptera were the predominant prey in this interval, whereas those of smaller size corresponded to beetles and those of greater size corresponded to amphibians and some belostomid bugs.

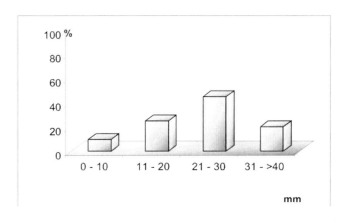

Rufescent tiger-heron

The size of prey ingested by Rufescent Tiger-Heron was uniformly distributed among the four intervals considered. Colubridae, amphibians and Pisces were the prey largest and insects were the prey smallest.

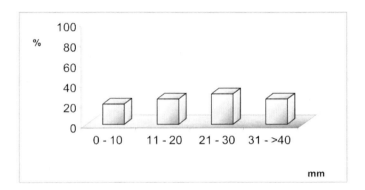

Whistling heron

Most of the prey ingested by Whistling Heron were within the ranges 0-20mm. (45%) and 21-40mm. (35%) and were predominantly insects, especially Orthoptera, Odonata and Hemiptera. While prey bigger with significantly lower percentages, 41-60mm. (15%) and 60 - > 80mm. (5%) were represented by species of Colubridae and amphibians.

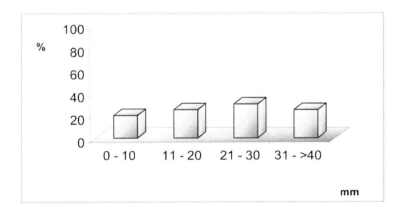

Snowy egret

The size of the prey of the Snowy Egret were mostly medium-sized belonging to the interval 11-20 mm, representing 45% of the total, including *H. Pequira, O. Paraguayensis* and *A. Combrae*. A 25% were prey whose size ranged from 21-30mm. and 20% had less than 10mm in size, both groups represented by fish and insects. Larger prey represented only 10% and included fish such as juveniles of *H. Malabaricus, P. albicans* and *E. Virescens*.

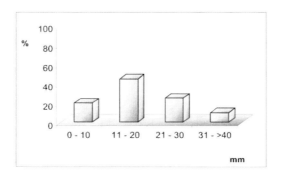

Stripe-backed bittern

The largest percentage of prey (55%) of Stripe-Backed Bittern were represented by insects, within the class interval 31-40 mm. Among this prey Orthoptera and Odonata were the most important, while a smaller percentage (10, 15 and 20%) correspond to Coleoptera (Curculionidae, Dytiscidae and Hudrophilidae) and some not Arachnida identified.

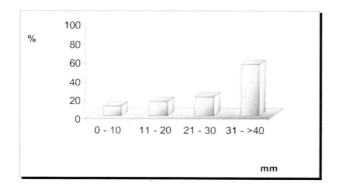

Great egret

The size of the largest of the prey (40%) corresponded to the interval 21-30 mm., the majority were fish such as *A. rubripinnis, A. fasciatus, S. spilopleura, C. platanus*, while also the fish with insects were smaller intake (0-10 and 11-20 mm.).

A 15% of the prey exceeded the 30 mm. and was represented by juveniles of *Salminus maxillosos, Leporinus obtusiden* and within the Decapoda by *Trichodactyllus borelianus*.

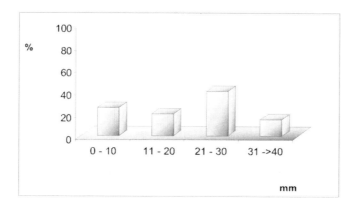

Black-crowned night-heron

The size of the highest percentage of prey in the Black-Crowned Night-Heron, corresponded to the highest class interval (61 -> 80 mm.) and were mainly represented by *S. marmoratus* and *L. obtusidens*. Intakes within intervals means, 21-40 mm. and 41-60 mm., both with 25% of the total, were smaller fish and crustaceans, while the smaller prey, 0-20 mm. (10 %) were insects.

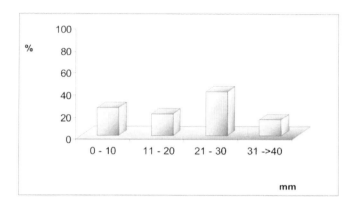

DIETARY SELECTIVITY

There were no significant results applying the Spearman's rank correlation coefficient, according to the following detail:

Cocoi Heron = rs = 0,078 P = > 0,001
Striated Heron = rs = 0,94 P = > 0,001
Cattle Egret = rs = 0,84 P = > 0,001
Rufescent Tiger-Heron = rs = 0,87 P = > 0,001
Whistling Heron = rs = 0,76 P = > 0,001
Snowy Egret = rs = 0,42 P = > 0,001
Stripe-Backed Bittern = rs = 0,56 = = > 0,001
Great Egret = rs = 0,072 P = > 0,001
Black-Crowned Night-Heron = rs = 0,41 P = > 0,001

RHYTHM OF THE FEEDING ACTIVITY

For most species primarily diurnal feeding activity pattern was distinguished, being able to differentiate two models: *bell model*, with a characteristic peak at noon and a reduction at dusk and the *bimodal model* representing two peaks of trophic activity.

According to the IF values obtained, Cocoi Heron, Striated Heron, Cattle Egret and Rufescent Tiger-Heron and Whistling Heron respond to the first of them, however in the activity of this latter species the peak of activity was not as marked.

Snowy Egret and Stripe-Backed Bittern respond to the second model, with increased activity in the early hours and afternoon with a lower activity towards noon and sunset.

In the case of Great Egret while viewing a peak of activity in mid-day hours, there is a marked positive trend from 05:00 pm. In this regard, although no studies were conducted during the night, can be noted for this species has also a nocturnal feeding.

Black-Crowned Night-Heron responds to a decreasing linear model, representing a twilight and nocturnal trophic activity. Full stomachs were found early in the morning (06:00-07:00 am.) while more digested contents were found as the time went by, finding empty stomachs at noon hours.

Rate of feeding activity for each specie calculated as the average satiety index (IF) for each time interval of capture.

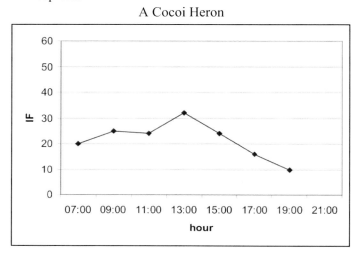

A Cocoi Heron

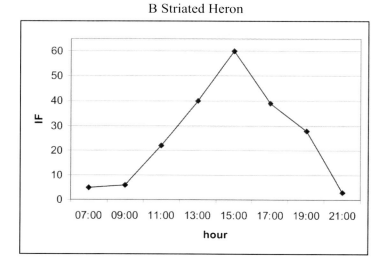

B Striated Heron

C Cattle Egret

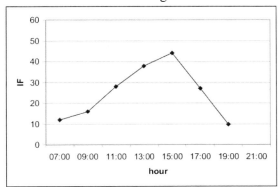

D Rufescent Tiger-Heron

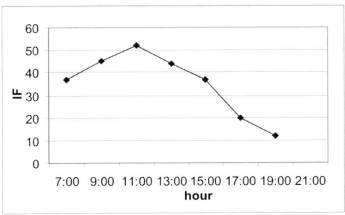

E Whistling Heron

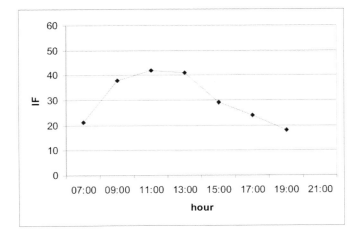

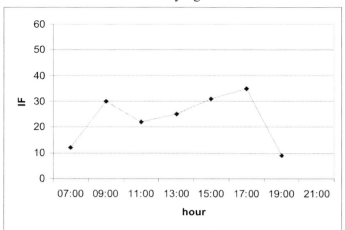

F Snowy Egret

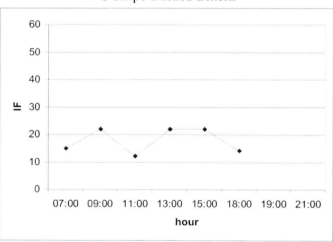

G Stripe-Backed Bittern

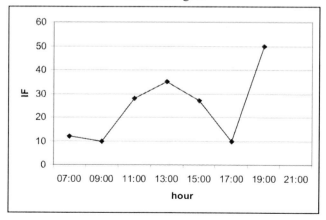

H Great Egret

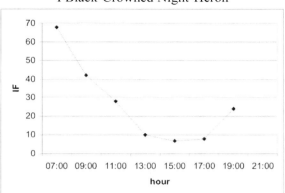

HABITAT PREFERENCE

All species used more than one "UVEs", Stripe-Backed Bittern being which was associated with less environmental units (only two).

Cocoi Heron and Great Egret were the species that used most environments, being "pastures", the unique environment to which was not associated none of them.

Only two species, Cocoi Heron and Black-Crowned Night-Heron, showed preference for more of one "UVES".

The unit "open water" was the most used, being the Whistling Heron the only species that was not associated with it, followed by "aquatic vegetation" which was not used for that species or Stripe-Backed Bittern.

The environment associated with a lower number of species was "pastures", only two species were observed using this.

Especie	open waters	aquatic vegetation	gallery forests	grasslands	pastures	forest	beach
Cocoi Heron	0.31	0.4	0.06	0.13	-	0.09	0.01
Striated Heron	0.11	0.51	0.2	-	-	-	-
Cattle Egret	0.13	0.08	-	-	0.34	-	-
Rufescent Tiger-Heron	0.2	0.37	0.18	-	-	-	-
Whistling Heron	-	-	0.18	-	0.46	0.28	-
Snowy Egret	0.1	0.46	-	0.2	-	0.1	0.17
Stripe-Backed Bittern	0.08	-	-	0.43	-	-	-
Great Egret	0.3	0.48	0.1	0.2	-	0.19	0.12
Black-Crowned Night-Heron	0.11	0.38	0.21	0.62	-	-	0.14

TROPHIC RELATIONSHIPS

Trophic relationships analysis, that provides the relative position of each species within the community, was calculated based on the basic data matrix and the values obtained by the application of the Relative Importance Index. Each heron studied species represented a taxonomic operational unit (TOU). Codified characters were:

1. Percentage frequency of occurrence.
2. Percentage numerical.
3. Percentage by volume.

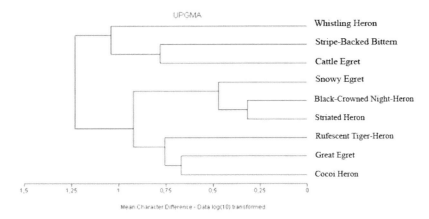

The phenogram showed the existence of well differentiated groups. The group of the mainly ichthyophagous herons can be easily noticed, and at the same time, it presented to intra-groups related among each other by the corporal size of the included species. Large bodied species (*Ardea cocoi, Casmerodius albus* and *Tigrisoma lineatum*) are clearly separated from the medium bodied ones (*Butorides striatus, Nycticorax nycticorax* and *Egretta thula*). In a separate group, the species which are not closely related to the water bodies and whose basic food items are insects (*Bubulcus ibis, Ixobrychus involucris* and *Syrigma sibilatrix*) were found.

DISCUSSION

The coexistence of ecologically similar species within an environment such as the Parana River floodplain occurs along evolutive paths leading species to differentiate themselves in their use of resources (Schoener, 1974). Diets and feeding habits constitute outstanding alternatives (Kushlan et al, 1982; Kushlan, 1983; Hancock, J. and Kushlan, 1984; Custer and Peterson, 1991)

Concerning trophic spectra of the species studied, the results of the index of relative importance (IRI) reveal that fish constitute the basic category, as it happens in particular with *Ardea cocoi, Butorides striatus, Tigrisoma lineatum, Casmerodius albus, Egretta thula* and *Nycticorax nycticorax*. The remaining items fall within the secondary or accessory categories, depending on the species. Insects, arachnids, crustaceans, amphibians, reptiles and mammals are found at these levels. In the case of *Bubulcus ibis, Syrigma sibilatrix* and *Ixobrychus involucris* the basic category is constituted by insects, while fish, amphibians, arachnids and crustaceans are found within the secondary or accessory categories (Beltzer, 1995; Zacagnini and Beltzer and 1982).

Body size and beak length show significant differences concerning the range of prey size and class intervals with greater percentage of incorporation. In the case of herons of larger size, they capture bigger prey more frequently during the intervals. Although species of larger size such as *Ardea cocoi*, *Casmerodius albus* and *Tigrisoma lineatum* present some small feeding items, it is doubted whether these have been incorporated intentionally by the birds or if they corresponded to intakes performed by the preys the birds captured.

Considering diet overlapping per pair of herons which has allowed determining their similarity and overlapping, it becomes evident that heron species establish a clear isolation mechanism in the trophic dimension of the niche, which favors their coexistence. Some authors (Andrewartha and Birch, 1954; Bowman, 1961) put forward the idea of disregarding competition as an important factor in the structure of communities. However, Begon et al. (1988) believe that interspecific competition can play a key and powerful role in the organization of communities.

The results of Schoener (1983) and Connel (1983) seem to indicate that interspecific competition is a widespread phenomenon, but its percentage of occurrence among species is relatively occasional. For this case, herons would integrate a system of competition by exploitation (Elton and Miller, 1954; Miller, 1967) in which they would use a resource without reducing the possibility for others to exploit it as well. In the words of Wiens and Rotenberry (1979), Jaksic et al. (1981), overlapping in itself does not provide evidence of competition in a natural environment but it constitutes, as it has already been indicated, probability of encounter (Colwell and Futuyma, 1971).

The greatest concentrations of herons were found in the area of study at the limnophase. During this hydrological status small shallow water bodies which during the potamophase were connected to the river or to permanent lenitic bodies, offer herons the possibility to exploit fish that have been trapped and unless used as food, are bound to die of desiccation. In the Parana River, the amplitude of the floodplain ensures the availability of lenitic water bodies all along the cycle, as it is the case of Carabajal Island where this study was conducted. The amplitude of the trophic niche and the standardized values clearly evidence that the fundamental or effective niche would seem to express itself seasonally and this has a tight linkage with the offer in each season of the year, aspect that reveals the great plasticity of the birds living in an area characterized by the marked variations in the hydrometric level along the annual cycle (Beltzer and Neiff, 1992).

The adjustment of the studied herons to the environment is expressed by means of the values of alimentary efficiency obtained (Ricklefs, 1996). The success of each organism depends on their adaptations. Foods of animal origin are more easily digested; therefore the values of the efficiencies must range between 60 and 90%, being the latter the value corresponding to carnivores (Ricklefs, 1996). In the present study, values close to 90% were recorded, coincidentally with the carnivore habits of the total species studied.

The non-significant values of the dietary selectivity coincide with that exposed by Jenni (1973), who states that herons are in general opportunistic birds regarding food, so that their spectra may vary widely according to the seasons, general conditions and offer of resources. According to Bozinovic and Merrit (1991), and Foley and Cork (1992) this adaptation to environmental variability evidences the plasticity of the organisms which is expressed by the behavioral, anatomic and physiological mechanisms.

As to Wilson's classification (1980), the large units of vegetation and environment (GUVA, according to their initials in Spanish), would represent the core area (zone which is more frequently used within the area of influence or area of activity) for the herons which are basically ichthyophagous, whereas grasslands are core areas for herons, that are basically insectivorous. The extraordinary richness of the fauna and aquatic vegetation allows a greater food supply, favoring an increase in the survival rate of potential prey (Bell, et al., 1991).

Due to the characteristics of its circadian rhythm, *N. nycticorax* would establish a clear isolation mechanism in the temporal dimension of the niche with the remaining studied species. Its nocturnal habits agree with that set forth by Beltzer and Oliveros (pers. obs.) in the area of study and with the specific literature Hudson, 1974; Hanzak, 1968; Watmough, 1978; Saver, 1984).

With regard to the remaining species, the observations and captures that enable the assessment of the rhythm of the feeding behavior were conducted from dawn to dusk. This is a fact worth highlighting since some species which were considered of diurnal activity also feed at nighttime. Robert, et al. (1989), when referring to *N. nycticorax,* mentions a frequency of diurnal occurrence of 93% and a nocturnal one of 38.2 %; while for the herons of the genus Egretta, the values were 64.8% and 7.8% respectively. On the other hand, Scortecci (1969) observes diurnal and nocturnal feeding activity for the *C. albus*. With the exception of the great white heron, the small white heron and the sunflower heron, the others respond to a "bell" model that indicates a peak of activity towards the hours of noon, while the species already mentioned present a bimodal tendency, which means two peaks of activity.

The values for habitat selection express a marked preference, in the case of the basically ichthyophagous birds, to the environment of aquatic vegetation whether floating or rooted. This GUVA constitutes the main source for obtaining resources for such herons. Arboreal formations, with seemingly lower values, represent the resting, sleeping and nesting area. On the other hand, the species for which insects constitute the basic diet express greater preference for the environment with grassland and low levels in relation to other units of vegetation and environment. This differential utilization of the habitat is a form of spatial segregation as pointed out by Schoener (1974), aspect discussed by Pianka (1975), who regards the trophic dimension as being the most important one.

CONCLUSIONS

The species of herons studied exhibit a patch-type distribution, unpredictable in time and space. Habitat offer is strongly conditioned by the pulsating hydrosedimentological regime, on whose characteristics spatial heterogeneity and as a consequence, species distribution in the landscape depends.

Considering what has been discussed, it is pointed out that the environments of the Parana River floodplain constitute a mosaic of favorable/unfavorable (GUVA) habitats, representing sites which are temporally accessible.

Isolation mechanisms are expressed in the presence of the contrasting conditions between imnophase and potamophase of the hydrological level. Therefore it becomes possible to define the optimal conditions for each ardeid along the slope of resources.

The trophic versatility of these herons is the consequence of a structural design which favors the adoption of different capture mechanisms, allowing them to adjust to the different types of prey, a fact that is evidenced by stomach contents that have shown a wide range of items.

The low diet overlapping is only an evidence of the encounter probability among the studied species. The recorded values were ostensibly low and nevertheless, they provide no evidence of competition. Therefore, these results reveal an isolation mechanism in the trophic dimension of the niche, which is manifested in the flexibility of the capture techniques, body size and the environmental units used. This allows pointing out that, regardless of their similitude, they subtly exploit different resources with spatial (spatial dimension) and circadian (temporal dimension) differences. In brief, coexistence is based on the differential use of resources as a basic isolation mechanism and less manifest in space and time.

Even though they constitute a single guild, they show differences in the feeding patterns, differences in the alimentary spectra in terms of offer, an aspect that is associated to the fact that the effective niche is expressed seasonally. Thus, the trophic sub niche has the greatest weight in relation to the temporal and spatial dimensions, there being agreement between feeding ecology and the design or adaptative series of such species.

Interaction among the different species of herons, as well as that observed in other species, makes the insular environments of the floodplain of the Parana River an object of study of particular interest to gain an insight into the aquatic system. Said research is considered valuable since it contributes to the preservation and management of biodiversity.

REFERENCES

Acosta Cruz, Torres, M. O. & Mugica Valdés, L. (1988). Sunicho trófico de *Dendrocygna bicolor* (Vieillot) (Aves: Anatidae) en dos arroceras de Cuba. *Ciencias Biológicas*, 19-20, 41-50

Amat, J. A. (1984). Las poblaciones de aves acuáticas en las lagunas andaluzas. Composición y diversidad durante un ciclo anual. *Ardeola, 31*, 61-79

Amat, J. A. & Soriguer, R. C. (1981). Alimentación primaveral de la garcita bueyera. *Doñana Acta Vertebrata. 8*, 207-213

Amat, J. A. & Aguilera, E. (1989). Some behavioural responses of Little Egret and Black-tailed Godwit to reduce prey lossos from kleptoparasites. *Ornis Scandinavica, 20(3),* 234-236

Andrehuarta, H. G. & Birch, L. C. (1954). The distribution and abundance of animals. *Univ. Chicago Press*, Chicago, 430.

Begon, M., Harper, H. J. & Towsend, C. R. (1988). Ecología. Individuos, Poblaciones y Comunidades. *Omega*, Barcelona, 866.

Bell, S. S., Mc Coy, E. D. & Mushinsky, H. R. (1991). Habitat structure. The physical arrangement of objects in space. *Chapman and Hall*, London, 438.

Beltzer, A. H. (1981). Nota sobre fidelidad y participación trófica de *Egretta alba egretta* (Gmelin, 1789) y *Egretta thula thula* (Molina, 1782) en ambientes del río Paraná medio (Ciconiiformes, Ardeidae). *Rev. Asoc. Cienc. Nat. Litoral,* 12, 136-139

Beltzer, A. H. (1983). c. Alimentación del benteveo (Pitangus sulphuratus) en el valle aluvial del río ParanA medio (Passeriformes: Tyrannidae). *Rev. Asoc. Cienc. Nat._Litoral, 14 (2)*

Beltzer, A. H. (1990). a. Biología alimentaria del gavilán común *Buteo magnirostris* (Aves: Accipitridae) en el valle aluvial del rio Paraná medio, Argentina. *Ornitol._Neotrop., 1(1),* 1-7

Beltzer, A. H. (1990). b. Biologia alimentaria del verdón común *Embernagra platensis* (Aves: Emberizidae) en el valle aluvial del río Paraná medio, Argentina. *Ornitol._Neotrop., 1(1),* 25-30

Beltzer, A. H. (1995). Las Ardeidae del valle de inundación del río Paraná. Nicho ecológico. Tesis de Maestría. *Universidad Nacional del litoral*, Santa Fe, 95.

Beltzer, A. H. (2003). Aspectos tróficos de la comunidad de aves de los esteros del Inerá. P. 257-272. En; Alabarez, B. B. (Eds.) Fauna del Iberá. *Universidad Nacional del Nordeste*, Corrientes.

Beltzer, A. H. & Oliveros, O. B. (1982). Alimentación del macá grande (Podiceps major) en el valle aluvial del río Paraná medio (Podicipediformes: Podicipedidae). *Rev. Asoc. Cienc. Nat. Litoral, 13,* 5-10

Beltzer, A. H. & Oliveros, O. B. (1987). Alimentación de los martín pescadores (*Ceryle torquata, Chloroceryle amazona* y *Chloroceryle americana*) en la llanura aluvial del río Paraná medio (Coraciiformes: Alcedinidae), *Ecología Argentina, 8,* 1-10

Beltzer, A. H. & Neiff, J. J. (1992). Distribución de las aves en el valle del río Paraná. relación con el regimen pulsátil y la vegetación. *Ambiente Subtropical, 2,* 77-102

Beltzer, A. H., Tomates, M. F. & Díaz, H. F. (2001). Nota sobre la dieta del favilán planeador *Circus buffoni* (Aves: Accipitridae) en el valle de inundación del río Paraná, Argentina. *Rev. FABICIB*, Fac. Bioq. Y Cs. Biol., UNL, Santa Fe, 5, 159-161

Beltzer, A. H., Quiroga, M. A. & Schnack, J. A. (2005). Algunas ardeidas del valle de inundación del río Paraná: consideraciones sobre el nicho ecológico y mecanismos de aislamiento. *INSUGEO*, Tucumán, Miscelánea, *14 (2),* 499-526

Bignal, E. M., Curtis, D. J. & Matthews, J. L. (1988). Islau land types. Bird habitats and nature conservation. Part. 1: Land use and birds on Islay. *NCC Chief Scientist Directorate*, 809.

Block, W. M., Brennan, L. A. & Gutierrez, R. J. (1992). Ecomorphological relationships of a guild of ground-foraging birds in northern California. *Oecologia, 87,* 449-458

Bozinovic, F. & Merrit, J. F. (1991). Conducta, estructura y función en micromamíferos en ambientes estacionales: mecanismos compensatorios. *Rev._Chil. Hist. Nat., 64,* 19-28

Bowman, (1961). R. I. Bowman, *Adaptation and differentiation of the Galapagos finches,* University of California Publications in Zoology 58 (1961), 1–302.

Brillouin, I. (1965). *Science and information theory*. Academic Press, New York, 346.

Buckelew, A. R. (1993). Grren-Backed Heron Swimming bejavior. *The Redstart, 108,* 59-98

Cabrera, A. (1932). La incompatibilidad ecológica. Una ley biológica interesante. *An. Soc. Cient. Arg., 114,* 243-260

Canevari, P., Blanco, D. E., Bucher, E. H., Castro, G. & Davidson, I. (1999). Los humedales de la ARgentina. Clasificación, Situación Actual, Conservación y Legislación. Publicación N 46, *Wetlands International*, 208.

Cody, M. L. (1974). *Competition and the structure bird communities*. Princeton Univ._Press, Princeton, 318.

Colwel, R. K. & D. J. Futuyma. (1971). On the measurement of niche breath and overlap. *Ecology, 52*, 567-576

Connel, J. H. (1983). On the prevalence and relative importance of interspecific competition, evidence from field experiments. *Am. Nat., 122*, 661-696

Cordiviola de Yuan, E. A. (1980). Campaña limnológica Keratella II en el río Paraná medio. Taxocenos de peces de ambientes leníticos. *Ecología*, 4, 103-113

Cowan, A. N. (1983). A modified penguin stomach tube. *Corella*, 7, 59-61.

Dodzhansky, T., Ayala, F., Stebbins, G. L. & Valentine, J. W. (1983). Evolución. *Omega*, Barcelona, 558.

Duncan, P. (1983). Determination of the use of habitat by horses in Mediterranean wetland. *J. Anim.Ecol., 52*, 93-109

Eckhardt, R. C. V. (1979). The adaptative syndromes of two guilds of insectivorous birds in the Colorado Rocky Mountains. *Ecol. Monogr., 49*, 129-149

Elton, C. & Miller, R. S. (1954). The ecological survey of animal communities: with a practical system of classifying habitats by structural characters. *Journal of Ecol., 42*, 460-496

Emison, W. B. (1968). Feeding preferences of the Adelie Penguin at Cape Crozier, Ross Island. Pp. 191-212 In *Antarctic bird studies* (Austin, I. O. Jr., Ed.) Antarct. Res. Ser., 12.

Erwin, R. M., Hatfield, J. S. & Link, W. A. (1991). Social foraging and feeding environment of the Black-crowned Night Heron in an industrialized estuary. *Bird Behav., 9*, 94-102

Ewins, P. J. & Hennessey, B. (1992). Great Blue Herons *Ardea herodias*, feeding at a fishing vessel offshore in Lake Erie. *Can. Field. Nat., 106:* 521-522

Faaborg, J. R. (1985). Trophic size structure of west Indian birds communities. *Proc._Natl. Acad. Sci., 79*, 1563-1567

Fernández Cruz, M. & Campos, F. (1993). The breeding of Grey Heron (*Ardea cinerea*) in western Spain: The influence of age. *Colonial Waterbirds, 16(1)*, 53-58

Frederick, P. C. & Collopy, N. W. (1989). Nesting success of five ciconiform species in relation to water conditions in the Florida Everglades. *Auk, 106*, 625-634

Frederick, P. C., Wyer, N. D., Fitzgerald, S. & Bennets, R. E. (1990). relative abundance and habitat preferences of Least Bitterns (*Ixobrychus exilis*) in the Everglades. *Florida Field Nat., 18(1)*, 1-9

Forbes, L. S. (1987). Feeding behaviour of Great Blue Heron at Creston, British Columbia. *Can.J.Zool, 65*, 3062-3067

Giller, P. S. (1984). Community structure and the niche. *Chapman & Hall*, London, 176.

Guillen, A., Prosper, J. & Echevarrias, J. L. (1994). Estimación de la dieta de la garcilla bueyera a partir del análisis de regurgitados de pollos: problemas debidos a la digestión diferencial de las presas. *Doñana Acta Vertebrata, 21(2)*, 204-210

Gómez-Tejedor, H. (1993). Comportamiento cleptoparásito en la garcita bueyera. *Butl.GCA, 10*, 71-73

Grinnel, J. (1917). The niche relationships of the California thraser. *Auk, 21*, 264-382

Hancock, J. & Kushlan, J. (1984). *The Herons Handbook.* New York City, NY: Harper and Row Publishers.

Hanzak, J. (1968). Gran enciclopedia ilustrada de las aves. *Lectura*, Caracas, 582.

Higgins, S. I., Coetzee, M. A. S., Marneweck, G. C. & Rogers, K. H. (1996). The Nyl river floodplain, South Africa, as a functional unit of the landscape: A revieu of corrents information. *Afr.J.Ecol., 34*, 131-145

Hudson, G. E. (1974). Aves del Plata. *Libros de Hispanoamérica*, Buenos Aires, 361.

Hurtubia, J. (1973). Trophic diversity measurement in sympatric species. *Ecology, 54 (4)*, 885-690

Hutchinson, G. E. (1979). El teatro ecológico y el drama evolutivo. *Blume,* Barcelona, 151.

Hutchinson, G. E. (1981). Introducción a la ecología de poblaciones. *Blume,* Ecología, Barcelona, 492.

Iriondo, M. & Drago, E. C. (1972). Descripción cuantitativa de dos unidades geomorfológicas de la llanura aluvial del Paraná medio, República Argentina. *Rev. Asoc. Geol. Arg., 27 (2),* 143-154

Iriondo, M. H., Paggi, J. C. & Parma, M. J. (2007). Introducción. Iriondo M. H. J. C. Paggi. & M. J. Parma (Eds.) The Middle Paraná River Limnology of a Subtropical Wetland. Springer-Verlag Berlin Heidelberg.

Jacksic, F. (1981). The guild structure of a community of predatory vertebrates in Central Chile. *Oecología, 49,* 21-28

Jenni, D. A. (1973). Regional variation in the food of Nestling Cattle Egrets. *Auk, 90 (4),* 821-826

Jordano, P. (1981). Alimentación y relaciones tróficas entre los passeriformes en paso otoñal por una localidad de Andalucía central. *Doñana Acta Vertebrata, 8,* 103-124

Junk, W. J., Bayley, P. B. & Sparks, R. E. (1989). The flood pulse concept in river floodplain systems. p: 110-127 in: Dodge, D. P.(Eds.) *Proceedings of the International Large River Symposium. Can. Spec. Publ. Fish. Aquat. Sci.,* 106

Kelly, J. P., Pratt, H. M. & Greene, P. L. (1993). The distribution, reproductive success, and habitat characteristics of heron and egret breeding colonies in the San Francisco Bay area. *Colonial Waterbirds 16(1),* 18-27

Kersten, B. M., Britton, R. H., Dugan, P. S. & Hafner, H. (1991). Flock feeding and food intake in Little Egrets: The effects of prey distribution and behaviour. *J. Anim. Ecol.,60,* 241-252

Kirkconnell, A., Garrido, O. H., Posada, R. S. & Cubillas, S. O. (1992). Los grupos tróficos de las aves cubanas. *Poeyana, 415,* 124

Kovach, W. L. (1999). *MVSP*. Multivariate Statical Package for IBM – PCm version 03.

Kushlan, J. A. (1976a). Wading bird predation in seasonally fluctuating pon. *Auk, 93,* 464-476

Kushlan, J. A. (1976b). Feeding behavior of North American Herons. *Auk, 93(1),* 86-94

Kushlan, J. A. (1978). Nonrigorous foraging by Robbing Egrets. *Ecology, 59(4),* 649-653

Kushlan, J. A. (1981). Resource use strategies of wading birds. *Wilson Bull., 93(2),* 145-163

Kushlan, J. A., Bass Jr. O. L., Loope, L. L., Robertson, Jr., W. B., Rosendahl, P. C. & Taylor, D. L. (1982). *Cape Sable sparrow management plan*. National Park Service, South Florida Research Center Report M-660. 37.

Kushlan, J. A. (1983). Pair formation behaviour of the Galapagos Lava Heron. *Wilson Bull. 95,* 118-121.

Landres, P. B. & MacMahon, J. A. (1980). Guilds and community organization: analysis of an oak woodlands in Sonora, México. *Auk, 97,* 351-365

Lekuona, J. M. & Campos, F. Le succes de reproduction du Héron cendré Ardea cinerez dans le bassin d'Arcachon. Alauda, *63(3),* 179-183

Levins, R. (1968). *Evoloution in changing environments*. Princeton Univ. Press, New Jersey, 120.

Lopez Ornat, A. & C. Ramo. (1992). Colonial waterbirds populations in the Sian Ka an Biosphere Reserve, Quintana Roo, México. *Wilson Bull., 104(3),* 501-515

Mc Neil, R., Drapeau, P. & Pierotti, R. (1993). Nocturnality in colonial waterbirds: occurrence special adaptations and suspected benefits. *Current Ornithol., 10,* 187-246

Maddock, M. & Geering, D. J. (1994). Range expansion and migration of the Cattle Egret. *Ostrich 65,* 191–203

Maitlan, P. S. (1978). Biology of freshwaters. *Blackie,* London, 244.

Magurran, A. E. (1989). La diversidad ecológica y su medición. *Vedra,* Barcelona, 199.

Martínez, M. (1993). Las aves y la limnología. *Conferencias de Limnología,* La Plata. 127-140

Marquis, M. & Leitch, A. F. (1990). The diet of Grey Heron *Ardea cinerea* breeding at Loch Leven, Scotland and the importance of their predation on ducklibgs. *Ibis, 132,* 335-549

Miller, R. S. (1967). Pttern and proccesd in competition. *Ecol. Res., 4,* 1-74

Moreira, F. (1992). Aves piscivoras em ecossistemas estuarinos: a dieta da garca branca pequena *Egretta garzetta* e da garca real *Ardea cinerea* num banco de vasa do stuario do Tejo. *Airo, 3(1),* 9-12

Morrone, J. J. (2001). Biogeografía de América Latina y el Caribe. *Cites, UNESCO,* Sea, Zaragoza, 148.

Neiff, J. J. (1975). Fluctuaciones anuales en la composición fitocenótica y biomasa de la hidrofitia en lagunas isleñas del Paraná medio. *Ecosur, 2(4),* 153-183

Neiff, J. J. (1979). Fluctuaciones de la vegetación acuática en ambientes del valle de inundación del Paraná medio. *Physis, Sec. B, 38(95),* 41-43

Neiff, J. J. (1986a). Las grandes unidades de vegetación y los ambientes insulares del río Paraná en su tramo Candelaria-Itá Ibaté. *Rev. Asoc. Cienc. Nat. Litoral, 17(1),* 7-30

Neiff, J. J. (1986b). Aspectos metodológicos y conceptuales para el conocimiento de las áreas anegables del Chaco Oriental. *Ambiente Subtropical, 1,* 1-4

Neiff, J. J. (1990). Ideas para la interpretación ecológica del Paraná. *Interciencia, 15(6),* 424-441

Neiff, J. J. (1999). El regimen de los pulsos en ríos y grandes humedales del Sudamérica. p. 97-146. En: Malvarez, A. I. & P. Kandus. Tópicos sobre grandes humedales Sudamericanos. *ORCYT-MAB (UNESCO),* Montevideo, 224.

Nudd, T. D. (1983). Niche dynamics and organization of waterfowl guild in variable environments. *Ecology, 64,* 319-330

O'Connor, T. G. (1993). The diet of nestling Cattle Egrets in the Transvaal. *Ostrich, 64,* 4-45

Orians, G. H. (1969). The number of bird species in some tropical forest. *Ecology, 50(5),* 783-796

Osborne, D. R., Beissinger, S. R. & Bourne, G. R. (1983). Water as an enhancing factor in bird community structure. *Caribb.J.Sci., 19(1),* 35-38

Peris, S. J., Briz, F. J. & Campos, F. (1994a). Recent changes in the food of the Grey Heron *Ardea cinerea* in Central West Spain. *Ibis, 136(4),* 488-489

Peris, S. J., Briz, F. J. & Campos, F. (1994b). Shifts in the diet of the Grey Heron (*Ardea cinerea*) in the Duero Basin, Central-West Spain, following the introduction of exotic fish species. *Folia Zool., 44(2),* 97-102

Pianka, E. R. (1973). The structure of lizards communities. *Am. Rev. Ecol. Syst., 4,* 53-74

Pianka, E. R. (1975). Niche relationships of desert lizards. En: M. L. Cody & J. M. Diamons(Eds.) *Ecology and Evoloution of Communities.* Harvard Univ. Press, 292-314

Pianka, E. R. (1982). Ecología evolutiva. *Omega*, Barcelona, 365.

Pinkas, L., Oliphant, M. S. & Iverson, Z. I. (1971). Food and feeding habits of albacore bluefin tuna and bonito in the California waters. *Dep. of Fish and Game. Fish Bull., 150*, 1-105

Ramo, C. & Busto, C. (1993). Resource use by herons in a Yucatán wetland during the breeding season. *Wilson Bull., 105(4),* 573-586

Remmert, H. (1988). Ecología. Autoecología. Ecología de poblaciones y esudios de ecosistemas. *Blume*, Barcelona, 304.

Robert, M., Mc Neil, R. & Leduc, A. (1989). Conditions and significance of night feeding in shorebirds and other water birds in a tropical lagoon. *Auk, 106*, 94-101

Rohwer, S. (1988). Foraging differences between White and Dark morphs of the Pacific Reef Heron *Egretta sacra. Ibis, 132*, 21-26

Root, O. M. (1968). Spizella pallida. Pages 1186-1208 in A. C. Bent, editor. Life histories of North American cardinals, grosbeaks, buntings, towhees, finches, sparrows, and allies. U.S. National Museum Bulletin 237.

Saver, F. (1984). Aves acuáticas. *Blume*, Barcelona, 286.

Shealer, D. A. & S. W. Kress. (1991). Nocturnal abandonment response to Black-Crowned Night Heron disturbance in a Common Tern colony. *Colonial Waterbirds, 14(1),* 51-56

Schefler, W. C. (1981). Bioestadística. *Fondo Educativo Interamericano*, México, 267.

Schoener, T. W. (1974). Resource partitioning in ecological communities. *Science, 185*, 27-39

Schoener, T. W. (1983). Field experiments on interspecific competition. *Am. Nat.*, 122, 240-285

Sokal, R. R. & Rohlf, F. J. (1979). Biometría. Principios y métodos estadísticos en la investigación biológica. *Blume*, Madrid, 832.

Tosi, G. & Toso, S. (1979). Night Herons *Nycticorax nycticorax* wintring in the po River Valley. *Ibis, 121*, 336

Walker, Y. (1987). Compartimentalization and niche defferentiation: causal patterns of competition and coexistences. *Acta Biotheor., 36*, 215-239

Watmough, B. R. (1978). Observation on nocturnal feeding by Night Heron *Nycticorax nycticorax. Ibis, 120*, 356-358

Wiens, J. A. & Rotenberry, J. (1979). Diet niche relationship among North American grassland and shrubteppe birds. *Oecologia, 42*, 253-292

Wilson, M. F. (1974). Avian community organization and habitat structure. *Ecology, 55*, 1017-1029

Wilson, E. O. (1980). Sociobiología. La nueva síntesis. *Omega*, Barcelona, 701.

Whitfield, A. K. & Cyrus, D. P. (1978). "Feeding succession and zonation of aquatic birds at False Bay, Lake St Lucia" *Ostrich 49*, 8-15.

Wolf, B. O. & Jones, S. L. (1989). Great Blue Heron death caused by predation on Pacific Lamprey. *Condor, 91*, 482-484

Zaccagnini, M. E. & Beltzer, A. H. (1982). Alimentación de *Bubulcus ibis ibis* L. 1758 y su relación trófica con *Egretta thula thula* (Molina, 1782) el Leales, Tucumán (Ciconiiformes: Ardeidae). *Rev. Asoc. Cienc. Nat. Litoral, 13*, 73-80

Ziswiler, V. (1980). Zoología especial vertebrados. T. 2. *Omega*, Barcelona, 413.

In: Trends in Ornithology Research
Editors: P. K. Ulrich and J. H. Willett, pp. 95-119

ISBN: 978-1-60876-454-9
© 2010 Nova Science Publishers, Inc.

Chapter 3

REFLECTIONS OF WINTER SEASON LARGE-SCALE CLIMATIC PHENOMENA AND LOCAL WEATHER CONDITIONS IN ABUNDANCE AND BREEDING FREQUENCY OF VOLE-EATING BIRDS OF PREY

Tapio Solonen

Luontotutkimus Solonen Oy, Neitsytsaarentie 7b B 147, FI-00960 Helsinki, Finland

ABSTRACT

I examined long-term (1986–2008) data on the number of occupied territories and breeding frequency (active nests) of nine species of vole-eating birds of prey in southernmost Finland, using generalized linear models. Explaining variables included wintertime and monthly large-scale climatic conditions indicated by North Atlantic Oscillation (NAO), mean winter and monthly mean ambient temperature and depth of snow cover at five local weather stations, as well as indices of autumn and spring abundance of voles at three localities within or near to the study area. The birds of prey included six site-tenacious species, of which four (*Bubo bubo, Glaucidium passerinum, Strix aluco, Strix uralensis*) were mainly sedentary and two (*Circus aeruginosus, Buteo buteo*) migratory ones, and three more or less nomadic species (*Falco tinnunculus, Asio otus, Aegolius funereus*). I expected that climatic effects were expressed in the numbers and breeding performance of birds of prey largely via their effects on highly fluctuating vole populations. In accordance with earlier findings, numbers and breeding of vole-eaters were largely governed by the abundance of small voles, confirming the suitability of my data to the present purpose. Large-scale climatic phenomena, indicating mild winter conditions, presented a nearly significant positive influence on the numbers and breeding frequency of southerly distributed site-tenacious species (*Buteo buteo, Bubo bubo, Strix aluco*). The combined effect of vole abundance and local mean winter temperature was negative both in sedentary *Strix aluco* and nomadic *Falco tinnunculus*. High temperatures in the beginning and at the end of winter showed positive associations. Thick snow cover combined with vole abundance showed positive associations with numbers and breeding frequency of various kinds of vole-eating birds of prey. The results followed largely my expectations though the link via vole abundance was inadequately demonstrated. My results suggest that the effects of global warming on various vole-eating birds of prey at high latitudes were both positive and negative, in particular due to mild winters. This would lead to changes in local populations and distribution ranges of species. Due to their flexible moving habits, nomadic species might be less seriously affected than site-tenacious ones that are more dependent on local resources, such as nest sites. From a local point of view and during a short period of time, however, the impact seemed to be more pronounced on nomadic species due to their sudden and drastic shifts.

INTRODUCTION

Global warming and its impact on the living conditions of various organisms have recently been considered frequently in a wide range of studies (e.g., Vitousek 1994, Hughes 2000, McCarty 2001, Zachos *et al.* 2001, Stenseth *et al.* 2002, Walther *et al.* 2002, Møller *et al.* 2004, Watkinson *et al.* 2004, Schwartz *et al.* 2006, Solonen 2008). Many of them have investigated effects of climate on the phenology, abundance, and various reproductive parameters of populations, and significant relationships have emerged. Large-scale climate indices have been suggested to predict ecological processes better than local weather variables (Forchhammer & Post 2004, Hallett *et al.* 2004), or on the contrary (Nielsen & Møller 2006). Often the most drastic climatic effects have been shown to focus on animal populations during winter or in the beginning of the breeding season, and they have concerned, in one way or another, the availability of food.

The close relationship between the fluctuating abundance of small voles and the occurrence and breeding success of vole-eaters is well known (e.g., Linkola & Myllymäki 1969, Mikkola 1983, Korpimäki 1985, Hanski *et al.* 1991, 2001, Solonen 2004, 2005). Small voles are the main prey of various avian and mammalian predators (e.g., Henttonen *et al.* 1987, Korpimäki & Norrdahl 1989, Hanski *et al.* 1991, Pucek *et al.* 1993, Hansson 1999, Ims & Andreassen 2000, Sundell *et al.* 2004). In northern Europe, their abundance shows wide cyclic oscillations that are, in general, most marked in the north, become progressively less pronounced southwards, and range from high-amplitude cycles of five years to low non-cyclic annual variations (Krebs & Myers 1974, Hansson & Henttonen 1985, Hanski *et al.* 1991, Norrdahl 1995, Stenseth 1999, Hansson *et al.* 2000, Sundell *et al.* 2004). In various localities, however, the previous regularity of the vole cycles have decreased (Lindström & Hörnfeldt 1994, Hanski & Henttonen 1996, Steen *et al.* 1996, Hansson 1999, Henttonen 2000, Laaksonen *et al.* 2002, Strann *et al.* 2002, Solonen 2004, Hipkiss *et al.* 2008, Ims *et al.* 2008). In northern Sweden, an important feature of the persistent decline in density and amplitude for vole species is a clear decrease in wintering success (Hörnfeldt 2004, Hörnfeldt *et al.* 2005). Potential causal factors behind these kinds of change include less favourable climatic conditions (Hörnfeldt 2004, Solonen 2004, Korslund & Steen 2006) and increased predation (Lindström & Hörnfeldt 1994, Hörnfeldt *et al.* 2005). Observations on the southern coast of Finland have suggested that the formerly regular three-year vole cycles were fading out, and that the general abundance of voles was declining (Solonen 2004). Coastal and inland vole abundances seem to fluctuate in significant synchrony, but their general levels in good vole years are considerably higher in inland rather than in coastal populations (Solonen 2004, Sundell *et al.* 2004). Spring densities have been declining and the highest peaks of cycles have been levelling off (Solonen 2004).

Density-dependent factors such as competition for food and predation, as well as density-independent climatic conditions govern the winter mortality of small voles (e.g., Hansen *et al.* 1999, Merritt *et al.* 2001, Aars & Ims 2002, Stenseth *et al.* 2002, Hörnfeldt 2004). The winter decline generally accelerates with increasing winter length and severity (Hansson & Henttonen 1985, Aars & Ims 2002), primarily due to food shortages (Stenseth *et al.* 2002, Huitu *et al.* 2003) or predation (Hanski *et al.* 2001, Korpimäki *et al.* 2002). In southernmost Finland, the winter decline in density is not significantly steeper in the coastal compared to the inland vole populations, but there are significant differences between the local populations of the southern coast (Solonen 2006). Small mammals seemed to overwinter more successfully in forests than in fields, and the negative effects of mild winters seemed to impact populations of field habitats. The direct contribution of high ambient temperatures on the overwinter population change seemed to be largely restraining, particularly in the depth of winter.

Prolonged unbroken periods of frost and snow are suggested to have more negative impact on animals than several shorter periods interspersed with mild spells (e.g., Newton 1998), but the opposite is also acknowledged. The latter implies that, for wintering small mammals, severe winters with thick snow cover, providing shelter from cold and predation, could be more favourable than mild ones (Hansson & Henttonen 1985, Merritt 1985, Sonerud 1986, Nybo & Sonerud 1990, Jędrzejewski & Jędrzejewska 1993, Pucek *et al.* 1993, Lindström & Hörnfeldt 1994, Merritt *et al.* 2001, Hörnfeldt 2004). In mild winters, the fluctuation of temperatures around the freezing point may be especially harmful by alternately freezing and thawing and thus wetting the wintering microhabitats of small mammals ("frost seesaw effect", Solonen 2004; cf. also Merritt 1985). Lethal effects may be cumulative, such as those causing deprivation of food and leading to death by starvation (Aars & Ims 2002, Korslund & Steen 2006), or short-cut ones, that kill fast by some physical way (e.g., freezing or drowning).

Near the southern coast of Finland, mild winters have predominated since the very cold one in 1987, coinciding with a lower level of vole abundance (Solonen 2004). Mild weathers are generally expected to be favourable for birds wintering at northern latitudes (e.g., Linkola & Myllymäki 1969, Hildén 1989). However, because the fluctuation of temperatures around the freezing point may be problematic for voles, it may consequently have adverse effects on vole-eating birds of prey (Solonen 2004). In mild winters a lesser proportion of voles seemed to survive from autumn to spring, providing an explanation for the negative effects of mild winters on vole-eating predators (Solonen 2004, 2006). Mild winters and a short supply of voles seemed to have more pronounced negative effects on vole specialists than on facultative vole-eaters. In coastal region, spring vole abundance contributed significantly to the occurrence of vole-eating owls, while the inland populations of owls seem to be largely governed by the vole abundance of the preceding autumn.

Theoretical studies suggest that predators may be particularly susceptible to the effects of climate change due to the direct effects of climate on the distribution and abundance of prey populations (Kareiva *et al.* 1993). Only few empirical studies have, however, investigated how climate affects prey-predator interactions (Visser *et al.* 1998, Rueness *et al.* 2003, Solonen 2004, Nielsen & Møller 2006, Hipkiss *et al.* 2008). Present study aims to elucidate contributions of large-scale climatic phenomena and local weather conditions on numbers and breeding frequency of vole-eating birds of prey in southern Finland. Relationships with large-scale climatic indices probably characterize broadly those general environmental conditions that affect numbers and breeding of vole-eating birds of prey while single local weather variables might give more detailed clues of the processes involved. I studied what kind of weather characteristics and which phases of winter period seemed to be of particular importance for the annual number of territories and nesting attempts in each species considered. Based on the earlier findings, I expected that 1) the effects of mild winters should be shown in numbers and breeding frequency of vole-eating birds of prey largely as negative responses due to their adverse impact on vole populations. The consequent logical prediction is that 2) the weather-related variations in food supply should be reflected especially in those species that are most dependent on voles, i.e., vole specialists. I also expected that 3) global climatic factors are significantly involved and they may outweigh the relationships with single local weather variables.

STUDY AREA

The study area was located near the northern border of the hemiboreal zone (see Hämet-Ahti 1981) in southernmost Finland (Figure 1). The hemiboreal zone halfway between the temperate and subarctic boreal zones is characterized by mixed coniferous forests, relatively cold winters, and mild summers.

The climate type in southern Finland is a northern temperate climate (see Drebs *et al.* 2002). Winters of southern Finland (average temperature of day is below 0) are usually 4–5 months long, and the snow covers the land about 4 months of every year, and in the southern coast, it can melt many times during winter, and then come again. The coldest winter days of southern Finland are usually -20° C, and the warmest days of July and early August can be 25-30° C. Summers in the southern Finland last 4 months (from the mid of May to mid of September).

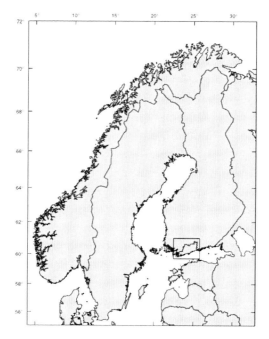

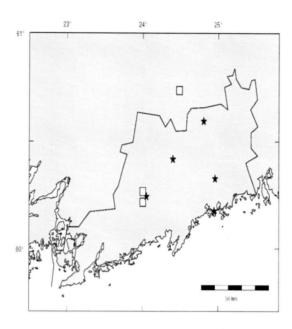

Figure 1. The study area of the Uusimaa region, southern Finland, in northern Europe. Locations of the vole trapping areas (squares) and weather stations (stars) are given.

MONITORING BIRDS OF PREY

My data are based on a nationwide monitoring program of Finnish birds of prey (Saurola 2008) and covers the years 1986–2008. I included six species of facultative vole-eaters and three species of nomadic vole specialists in the study (Table 1). For some species, our data were, however, scanty, and they are included only to examine if they show relationships resembling those of other species of similar kind of ecology. If they do not, it does not warrant any further-reaching conclusions.

Dependent Variables

The numbers of territories describe population size in each year and their difference between years may give clues of wintering success. The numbers of nests characterize breeding frequency in each species and year. Both of these variables were sufficiently available for five species (*Buteo buteo, Bubo bubo, Strix aluco, S. uralensis, Aegolius funereus*), for the rest only the number of territories was considered.

The numbers of *Bubo bubo, Buteo buteo* and *Aegolius funereus* showed significantly declining trends, and numbers of *Circus aeruginosus, Glaucidium passerinum* and *Falco tinnunculus* significantly increasing trends (Figure 2). Numbers of *Strix aluco, S. uralensis* and *Asio otus* fluctuated widely but showed no trends.

Table 1. The species of vole-eating birds of prey included in the present study. Food and migratory habits (e.g., Cramp 1980, Cramp & Simmons 1985)

Migratory facultative vole-eaters	
Western Marsh Harrier	*Circus aeruginosus* (Linnaeus 1758)
Common Buzzard	*Buteo buteo* (Linnaeus 1758)
Site-tenacious facultative vole-eaters	
Eurasian Eagle Owl	*Bubo bubo* (Linnaeus 1758)
Eurasian Pygmy Owl	*Glaucidium passerinum* (Linnaeus 1758)
Tawny Owl	*Strix aluco* Linnaeus 1758
Ural Owl	*Strix uralensis* Pallas 1771
Nomadic vole specialists	
Common Kestrel	*Falco tinnunculus* Linnaeus 1758
Long-eared Owl	*Asio otus* (Linnaeus 1758)
Tengmalm's Owl	*Aegolius funereus* (Linnaeus 1758)

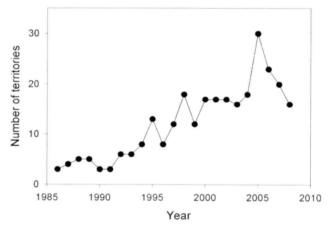

Circus aeruginosus

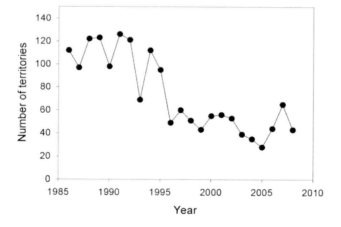

Buteo buteo

Figure 2. (Continued) See the legend below.

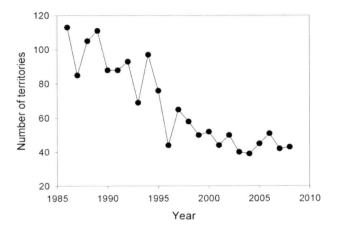

Bubo bubo

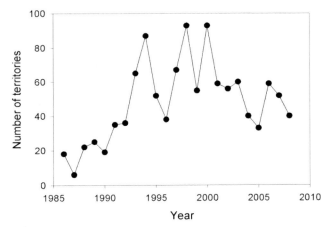

Glaucidium passerinum

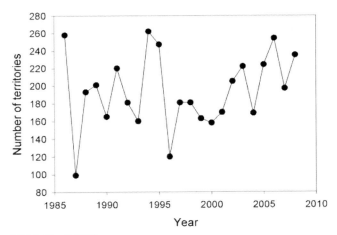

Strix aluco (t = 0.914, P = 0.371)

Figure 2. (Continued)

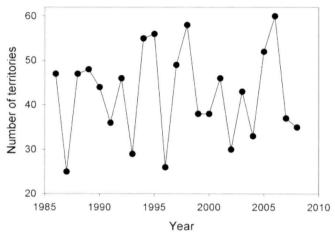

Strix uralensis

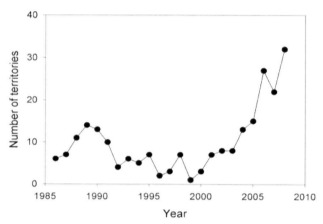

Falco tinnunculus

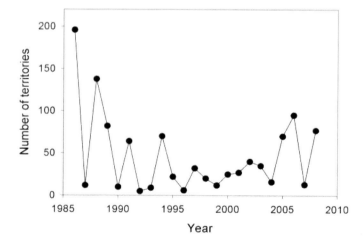

Asio otus

Figure 2. (Continued)

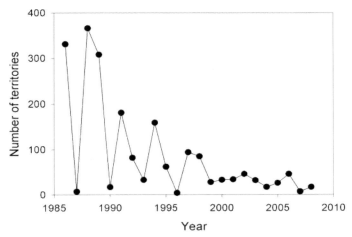

Aegolius funereus

Figure 2. Population trends of the vole-eating birds of prey in the present study area in 1986–2008: *Circus aeruginosus* (linear regression t = 9.048, P < 0.001), *Buteo buteo* (t = -7.106, P < 0.001), *Bubo bubo* (t = -9.100, P < 0.001), *Glaucidium passerinum* (t = 2.251, P = 0.035), *Strix aluco* (t = 0.914, P = 0.371), *Strix uralensis* (t = 0.105, P = 0.917), *Falco tinnunculus* (t = 2.652, P = 0.015), *Asio otus* (t = -1.208, P = 0.241), and *Aegolius funereus* (t = -3.630, P = 0.002).

TRAPPING OF SMALL MAMMALS

Small mammal assemblages were monitored biannually (twice a year) at three sites (Figure 1). Two of them were located in Uusimaa near the southern coast of Finland, situating 13 km apart (Lohja 60°16′N, 24°12′E and Kirkkonummi 60°13′N, 24°24′E) (1985–2008; Kimpari Bird Projects). Third trapping area, Loppi (60°43′N, 24°28′E) (1985–2005; A. Kaikusalo), was situated in southern Tavastia close to and within a landscape largely similar to the northern part of the present study area.

Snap trappings in Uusimaa were conducted each spring (early May, before the general breeding period of small mammals) and autumn (early October, before the overwintering period) at several standard points along four catching lines. At each of the four sites, there were used 16 points of three traps, the points situating about 25 m from each other, during two approximately 24-hr trapping periods, totalling 384 trap nights in each trapping. At Loppi, the trapping effort somewhat varied, averaging 247 trap nights per trapping.

The results (densities of small mammals) were expressed as catch indices (N), indicating the number of individuals caught per 100 trap nights. In the analyses, densities below the detection limit of the trapping method were replaced by a value of 0.001, corresponding to a density of about half of the minimum detected.

The species included in the vole indices used were the bank vole *Clethrionomys glareolus* (Schreber 1780) and field vole *Microtus agrestis* (Linnaeus 1761). The catch indices were considered to describe adequately the general levels of populations in the surroundings, because declines in catch indices in permanent sampling sites are not shown to be due to the repeated trappings there (Christensen & Hörnfeldt 2003). Both autumn and spring vole abundance fluctuated considerably (Figure 3) but there were no significant trends.

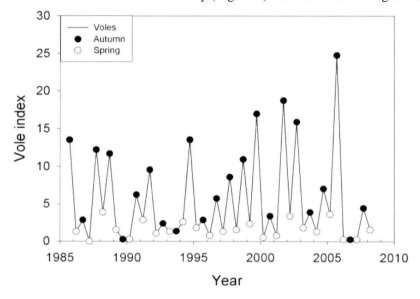

Figure 3. Fluctuations of preceding autumn (linear regression t = 0.616, P = 0.545) and spring (t = -0.126, P = 0.901) abundance of small voles in Lohja-Kirkkonummi area, southern Finland (Figure 1), in 1985–2008.

METEOROLOGICAL DATA

Large-scale Climatic Conditions

General large-scale climatic conditions were characterized by the North Atlantic Oscillation (NAO) indices (Jones *et al.* 1997), the positive values of which indicate milder and wetter winter weather in the North (Hurrell *et al.* 2001). The decline in the small mammal abundance during winter and the relationships between general winter weather conditions and vole-eating birds of prey were examined in the light of the winter NAO index (inclusive months December, January, February, and March) and monthly NAO indices of December, January, February, and March (http://www.cru.uea.ac.uk/cru/data/nao.htm). There were no significant correlations (r) between the monthly values of the NAO indices (P > 0.05). Winter NAO indices fluctuated but showed no trend (Figure 4).

Local Weather Data

Local weather variables were derived from five weather stations in Uusimaa (Helsinki-Vantaa airport, Helsinki, Lohja, Vihti, and Hyvinkää) (Figure 1). They included mean temperatures of winter months (Dec, Jan, Feb, Mar) and thickness of snow cover in the middle of each month (Finnish Meteorological Institute). Relationships of winter weather variables and the abundance of voles and vole-eaters were examined both with wintertime averages and separate monthly values.

Mean temperatures of the entire winter fluctuated without a trend (Figure 4). There was, however, a significant positive trend in December temperatures ($t = 2.079$, $P = 0.050$). Winter snow cover fluctuated considerably but there was no definite trend (Figure 5). There were, however, significant negative trends in snow cover of December ($t = -2.098$, $P = 0.048$) and January ($t = -2.316$, $P = 0.031$).

STATISTICAL PROCEDURES

Associations of climatic variables and vole abundance with numbers and breeding frequency of various species of vole-eating birds of prey were analyzed by linear generalized models. The analyses were performed by the statistical software package nlme (R Development Core Team 2008, Venables *et al.* 2008). I firstly considered the associations between the annual and monthly winter weather variables and the decline of small voles between the autumn and spring catches. Then the respective relationships between weather, voles, and the number of territories and nests of vole-eating birds of prey were examined. Both separate and combined effects of weather and food variables were considered. Best models in each case were selected primarily on the basis of Akaike Information Criterion (AIC) (see Burnham & Andersson 2002). If the AICs compared were close to each other (delta AIC less than 2), adjusted coefficients of determination (adj R^2) and the significance of variables included in the model served as selection criteria.

CLIMATIC EFFECTS ON SMALL VOLE POPULATIONS

There was a significant correlation between the vole indices of Lohja-Kirkkonummi area and Loppi both in autumn ($r = 0.538$, $P = 0.012$, df = 19) and in spring ($r = 0.500$, $P = 0.025$, df = 18). The general level of voles was, however, significantly higher at Loppi (autumn mean 17.58 ± 9.97 SD vs. 8.57 ± 6.55 SD, N = 23, $t_{42} = 3.57$, $P < 0.001$; and spring mean 9.97 ± 6.50 SD vs. 1.57 ± 1.11 SD, N = 20, $t_{41} = 6.10$, $P < 0.001$). In general, coastal and inland vole abundances seem to fluctuate in significant synchrony, but their general levels in good vole years are considerably higher in inland than in coastal populations (Solonen 2004, Sundell *et al.* 2004).

According to my predictions, effects of mild winters should be shown in numbers and breeding frequency of vole-eating birds of prey largely as negative responses due to their adverse impact on vole populations. However, I could not find any significant relationships between climatic variables used and the total values of spring abundance or winter decline of voles ($P > 0.05$). The general lack of significant relationships between climatic phenomena and spring vole abundance might be due to deficiencies in climatic variables to characterize significant features of climatic effects, or due to insufficient/ unrepresentative sample on voles.

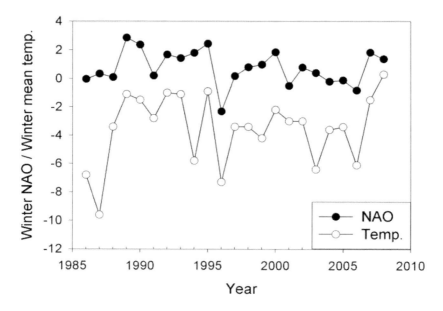

Figure 4. Fluctuations of winter NAO indices (linear regression t = -0.783, P = 0.442) and local winter mean temperatures (t = 0.898, P = 0.379) in Uusimaa, southern Finland (Figure 1), in 1985–2008.

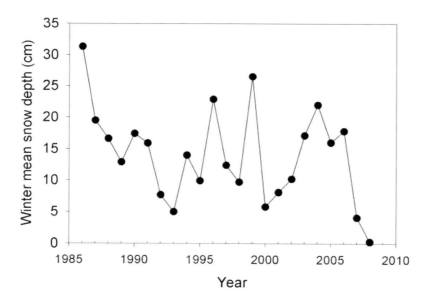

Figure 5. Fluctuations of winter mean snow depth (linear regression t = -1.750, P = 0.095) in Uusimaa, southern Finland (Figure 1), in 1985–2008.

Significant relationships were not found between weather variables and the vole indices in which the data of trapping localities as well as those of *Clethrionomys* and *Microtus* voles were combined. However, when the vole populations were considered separately, significant positive relationships emerged between the NAO indices and temperatures of February and March and the winter decline of *Microtus agrestis* at Loppi (Feb NAO, r = 0.472, P = 0.036, df = 18; Mar NAO, r = 0.553, P = 0.011, df = 18; Feb temp, r = 0.457, P = 0.043, df = 18;

Mar temp, r = 0.531, P = 0.016, df = 18). There were also nearly significant positive relationships (P < 0.10) between both March NAO index and snow cover of December and January, and winter decline of *Microtus agrestis* in the total data. In the Lohja-Kirkkonummi area, negative relationships have been found between NAO indices in December and January, and winter season population change in *Clethrionomys glareolus* and *Microtus agrestis,* respectively (Solonen & Ahola 2010; see also Solonen 2006).

According to earlier findings, the winter decline in density is not significantly steeper in coastal than in inland vole populations, but there are significant differences between local populations of the southern coast (Solonen 2006). Small mammals seemed to overwinter more successfully in forests than in fields, and the negative effects of mild winters seemed to impact the populations of field habitats. The direct contribution of high ambient temperatures on the overwinter population change seemed to be largely restraining, but especially so in the depth of winter. Potential causes behind these kinds of changes include, among others, less favourable climatic conditions (Hörnfeldt 2004, Solonen 2004). Climatic effects are expected to be especially pronounced after the overwintering period in spring (Hansen *et al.* 1999, Merritt *et al.* 2001, Aars & Ims 2002, Hörnfeldt *et al.* 2005, Solonen 2006). They are supposed to be largely due to such large-scale factors as the general global warming (Houghton *et al.* 1990) and the North Atlantic Oscillation that has recently brought warmer and wetter winters to Fennoscandia (Hurrell *et al.* 2001).

The three-year population cycles of small voles showed increasing irregularity during the latter half of the study period (Solonen & Ahola 2010). Contrary to *Microtus agrestis*, *Clethrionomys glareolus* showed an increasing trend in field habitats, and the fluctuation synchrony between the species disappeared. The positive relationships between the densities of species in successive trappings indicated the importance of some extrinsic factors (such as weather, habitat, or predation) in the dynamics of the populations. Benign weather conditions (indicated by high monthly NAO indices) in autumn contributed positively to the pre-overwintering densities of *Microtus agrestis,* while mild periods in the beginning and at the end of winter deepened the overwintering decline in *Clethrionomys glareolus* (Solonen & Ahola 2010).

Summarizing, my results, among some other studies, suggest that mild winters may have negative effects on vole populations by increasing their wintertime decline at least in *Microtus agrestis* and locally. This is in line with my assumptions, concerning the relationships between weather factors, vole abundance, and vole-eating birds of prey.

EFFECTS OF WINTER WEATHER CONDITIONS AND PREY ABUNDANCE ON VOLE-EATING BIRDS OF PREY

According to my predictions, effects of mild winters should be shown in the numbers and breeding frequency of vole-eating birds of prey as negative responses. This should consider, in particular, sedentary species that live in the environmental conditions of the breeding area throughout winter. However, weather-related effects on small voles should also be reflected heavily in those species that are most dependent on voles as food supply, i.e., the nomadic vole specialists.

Large-scale climatic phenomena indicated by winter NAO indices showed nearly significant positive relationships both in sedentary (*Bubo bubo, Strix aluco*) and migratory (*Buteo buteo*) species (Table 2). In nomadic *Falco tinnunculus* the combined effect of high winter NAO and preceding autumn vole abundance was, however, significantly negative. On monthly basis, significant positive relationships emerged with NAO indices of January (*Strix aluco*) and March (*Bubo bubo, Buteo buteo*), and negative associations with NAO of February (*Asio otus*) (Table 3).

Combined effect of mean winter temperature and vole abundance was negative both in sedentary *Strix aluco* and nomadic *Falco tinnunculus* (Table 4). Monthly mean temperatures both in the beginning (December) and at the end (March) of winter combined with vole abundance showed significant relationships with numbers of various species of vole-eaters (Table 5). High temperatures in December reflected in negative combined effect in nomadic *Aegolius funereus*, while combined effects of mean temperature of March and vole abundance were positive in sedentary *Bubo bubo* and migratory *Buteo buteo*.

Combined effects of thickness of whole winter snow cover and vole abundance were positive both in sedentary (*Bubo bubo, Strix aluco*) and nomadic (*Falco tinnunculus, Asio otus*) species (Table 6). The respective relationships were still more common with monthly values of snow cover, in particular that of December, and they concerned both sedentary, nomadic, and migratory species (Table 7). When considering single best models, snow cover of March seemed to be of most importance for *Strix aluco,* but the relationships were practically similar with the values of January and February as well.

Near the southern coast of Finland, spring vole abundance contributed significantly to the occurrence of owls, while in inland populations they seem to be largely governed by the vole abundance of the preceding autumn (Solonen 2004). The mild winter temperatures contributed negatively to the abundance of vole specialists. When the mildness of winters near the southern coast of Finland was characterized by the number of the days during which the ambient temperature at least once fell from plus to minus °C (intensity of the frost seesaw), it contributed negatively, in combination with the positive contribution of vole abundance, to the occurrence of nomadic vole specialists. In mild winters a lesser proportion of voles seemed to survive from autumn to spring, providing an explanation for the negative effects of mild winters on vole-eaters (Solonen 2004, 2006). The vole abundance may crash below the level needed for successful foraging (Solonen 2004, Ims *et al.* 2008). This suggests that in poor vole years voles were generally too scarce even without the effect of the frost seesaw to be effectively foraged by predators.

Table 2. Best models aiming to explain the relationships of winter NAO index, vole abundance and the number of territories and nests of various sedentary, nomadic and migratory vole-eating birds of prey in Uusimaa, southern Finland. Values of t and P for each variable included in the models as well as adjusted coefficient of determination (adj R²), F, df, and P for each model are given.

	Winter NAO	Autumn voles	Spring voles	NAO:voles	Adj R²	F	df	P
Territories								
Sedentary								
B. bubo	1.88 0.075		1.78 0.090		0.171	3.26	2,20	0.059
G. passerinum	0.59 0.565	1.04 0.311			-0.029	0.69	2,20	0.512
S. aluco	1.89 0.074	3.13 **0.006**		-1.53 0.143	0.284	3.91	3,19	**0.025**
S. uralensis	1.11 0.279	3.44 **0.003**			0.331	6.44	2,20	**0.007**
Nomadic								
F. tinnunculus	2.53 **0.020**	1.07 0.297		-2.95 **0.008**	0.216	3.02	3,19	0.055
A. otus	-0.71 0.487		2.20 **0.040**		0.134	2.71	2,20	0.091
A. funereus			2.45 **0.023**		0.185	5.98	1,21	**0.023**
Migratory								
B. buteo	1.97 0.063		1.08 0.293		0.118	2.47	2,20	0.110
C. aeruginosus	0.39 0.699	1.55 0.137		-1.23 0.235	0.051	1.40	3,19	0.275
Nests								
Sedentary								
B. bubo	1.32 0.202		2.82 **0.011**		0.254	4.75	2,20	**0.020**
S. aluco	2.03 0.056	3.42 **0.003**			0.379	7.70	2,20	**0.003**
S. uralensis		3.53 **0.002**			0.343	12.49	1,21	**0.002**
Nomadic								
A. funereus			2.02 0.056		0.123	4.09	1,21	0.056
Migratory								
B. buteo	1.75 0.095		1.58 0.131		0.134	2.70	2,20	0.092

Table 3. Best models aiming to explain the relationships between the NAO indices of winter months, vole abundance (a = preceding autumn, s = spring) and the number of territories and nests of various sedentary, nomadic and migratory vole-eating birds of prey in Uusimaa, southern Finland. Values of t and P for variables included in the models as well as adjusted coefficient of determination (adj R²), F, df, and P for each model are given.

	DecNAO	JanNAO	FebNAO	MarNAO	Voles	NAO:voles	Adj R²	F	df	P
Territories										
Sedentary										
B. bubo				2.78 **0.012**	1.57 0.131[s]		0.296	5.62	2,20	**0.012**
G. passerinum		1.31 0.204			1.75 0.096[a]	-2.07 0.052	0.108	1.89	3,19	0.165
S. aluco		2.51 **0.021**			2.99 **0.007**[a]		0.386	7.91	2,20	**0.003**
S. uralensis				2.00 0.060	3.76 **0.001**[a]		0.400	8.57	2,20	**0.002**
Nomadic										
F. tinnunculus		1.56 0.136			-1.53 0.142[s]		0.091	2.10	2,20	0.149
A. otus			-2.50 **0.021**		2.09 0.050[s]		0.324	6.28	2,20	**0.008**
A. funereus				-0.45 0.660	1.84 0.081[s]	1.50 0.150	0.286	3.94	3,19	**0.024**
Migratory										
B. buteo				2.34 **0.029**			0.169	5.48	1,21	**0.029**
C. aeruginosus				-1.75 0.095			0.086	3.06	1,21	0.095
Nests										
Sedentary										
B. bubo				-0.65 0.525	2.15 **0.045**[s]	2.12 0.048	0.436	6.67	3,19	**0.003**
S. aluco		3.33 **0.003**			3.69 **0.001**[a]		0.518	12.83	2,20	**0.000**
S. uralensis					3.53 **0.002**[a]		0.343	12.49	1,21	**0.002**
Nomadic										
A. funereus				1.32 0.201	1.91 0.071[s]		0.153	2.99	2,20	0.071
Migratory										
B. buteo				2.04 0.054	1.34 0.195[s]		0.174	3.31	2,20	0.057

Table 4. Best models aiming to explain the relationships of mean winter temperature, vole abundance and the number of territories and nests of various sedentary, nomadic and migratory vole-eating birds of prey in Uusimaa, southern Finland. Values of t and P for each variable included in the models as well as adjusted coefficient of determination (adj R²), F, df, and P for each model are given.

	Temperature	Autumn voles	Spring voles	Temp.:voles	Adj R²	F	df	P
Territories								
Sedentary								
B. bubo	-0.02 0.981		1.59 0.128		0.024	1.27	2,20	0.303
G. passerinum	0.75 0.461	1.08 0.291			-0.018	0.81	2,20	0.459
S. aluco	2.59 **0.018**	-0.45 0.661		-2.33 **0.031**	0.379	5.47	3,19	**0.007**
S. uralensis	0.89 0.386	3.44 **0.003**			0.316	6.09	2,20	**0.009**
Nomadic								
F. tinnunculus	3.34 **0.003**	-2.29 **0.034**		-3.01 **0.007**	0.289	3.98	3,19	**0.024**
A. otus	0.71 0.488		-0.50 0.625	-1.74 0.098	0.219	3.06	3,19	0.053
A. funereus			2.45 **0.023**		0.185	5.98	1,21	**0.023**
Migratory								
B. buteo	0.38 0.712		0.96 0.348		-0.046	0.51	2,20	0.606
C. aeruginosus	1.52 0.146	-0.59 0.564		-1.50 0.151	0.062	1.49	3,19	0.250
Nests								
Sedentary								
B. bubo			2.74 **0.012**		0.228	7.50	1,21	**0.012**
S. aluco	1.87 0.076	3.45 **0.003**			0.362	7.25	2,20	**0.004**
S. uralensis		3.53 **0.002**			0.343	12.49	1,21	**0.002**
Nomadic								
A. funereus			2.02 0.056		0.123	4.087	1,21	0.056
Migratory								
B. buteo			1.46 0.160		0.049	2.13	1,21	0.160

Table 5. Best models aiming to explain the relationships between the mean temperatures of winter months, vole abundance (a = preceding autumn, s = spring) and the number of territories and nests of various sedentary, nomadic and migratory vole-eating birds of prey in Uusimaa, southern Finland. Values of t and P for variables included in the models as well as adjusted coefficient of determination (adj R^2), F, df, and P for each model are given.

	Dec temp	Jan temp	Feb temp	Mar temp	Voles	Temp:voles	Adj R^2	F	df	P
Territories										
Sedentary										
B. bubo				-0.72 0.479	3.37 **0.003**[s]	2.64 **0.016**	0.312	4.32	3,19	**0.018**
G. passerinum		1.50 0.148					0.054	2.26	1,21	0.148
S. aluco	1.83 0.082				3.31 **0.004**[a]		0.308	5.88	2,20	**0.010**
S. uralensis		1.90 0.073			3.41 **0.003**[a]		0.398	8.26	2,20	**0.002**
Nomadic										
F. tinnunculus	2.58 0.018				-0.48 0.635[a]	-1.65 0.115	0.155	2.35	3,19	0.105
A. otus					2.24 0.036[s]		0.155	5.03	1,21	**0.036**
A. funereus	1.14 0.269				0.61 0.551[s]	-2.34 0.031	0.334	4.68	3,19	**0.013**
Migratory										
B. buteo				-0.18 0.863	2.72 **0.013**[s]	2.38 **0.028**	0.273	3.76	3,19	**0.028**
C. aeruginosus				0.63 0.537	-2.12 **0.048**[s]	-2.83 **0.011**	0.281	3.87	3,19	**0.026**
Nests										
Sedentary										
B. bubo				-0.71 0.489	5.01 **0.000**[s]	3.29 **0.004**	0.533	9.37	3,19	**0.001**
S. aluco		2.51 **0.021**			3.26 **0.004**[a]		0.430	9.28	2,20	**0.001**
S. uralensis		3.35 **0.003**		-2.05 0.054	3.39 **0.003**[a]		0.553	10.08	3,19	**0.000**
Nomadic										
A. funereus	1.07 0.298				0.23 **0.825**[s]	-2.35 **0.030**	0.293	4.03	3,19	**0.022**
Migratory										
B. buteo				0.14 0.888	3.24 **0.004**[s]	2.43 **0.025**	0.364	5.20	3,19	**0.009**

Table 6. Best models explaining the effects of mean winter snow cover and vole abundance on the number of territories and nests of various sedentary, nomadic, and migratory vole-eating birds of prey in Uusimaa, southern Finland. Values of t and P for each variable included in the models as well as adjusted coefficient of determination (adj R²), F, df, and P for each model are given.

	Winter snow depth	Autumn voles	Spring voles	Snow:voles	Adj R²	F	df	P
Territories								
Sedentary								
B. bubo	0.72 0.480		1.38 0.183		0.048	1.56	2,20	0.235
G. passerinum	-2.36 **0.029**	1.45 0.164			0.182	3.44	2,20	0.052
S. aluco	-3.26 **0.004**	-1.76 0.095		3.44 **0.003**	0.484	7.89	3,19	**0.001**
S. uralensis	-0.52 0.608	3.37 **0.003**			0.299	5.69	2,20	**0.011**
Nomadic								
F. tinnunculus	-4.26 **0.000**	-3.16 **0.005**		3.82 **0.001**	0.417	6.25	3,19	**0.004**
A. otus	-1.49 0.154		-1.87 0.077	3.26 **0.004**	0.446	6.89	3,19	**0.002**
A. funereus	-0.57 0.578		-0.34 0.740	1.50 0.151	0.222	3.10	3,19	0.052
Migratory								
B. buteo	0.13 0.896		0.87 0.392		-0.053	0.45	2,20	0.645
C. aeruginosus	-1.47 0.156	1.57 0.131			0.087	2.05	2,20	0.155
Nests								
Sedentary								
B. bubo	-1.18 0.254		-0.68 0.503	2.12 **0.048**	0.323	4.49	3,19	**0.015**
S. aluco	-1.64 0.117	0.02 0.987		1.43 0.169	0.309	4.28	3,19	**0.018**
S. uralensis	-1.37 0.186	3.76 0.001			0.369	7.44	2,20	**0.004**
Nomadic								
A. funereus	-0.58 0.568		-0.79 0.438	1.79 0.090	0.221	3.08	3,19	0.052
Migratory								
B. buteo			1.46 0.160		0.049	2.13	1,21	0.160

Table 7. Best models aiming to explain the relationships between the snow cover of winter months, vole abundance (a = preceding autumn, s = spring), and the number of territories and nests of various sedentary, nomadic and migratory vole-eating birds of prey in Uusimaa, southern Finland. Values of t and P for variables included in the models as well as adjusted coefficient of determination (adj R²), F, df, and P for each model are given.

	Dec snow	Jan snow	Feb snow	Mar snow	Voles	Snow:voles	Adj R²	F	df	P
Territories										
Sedentary										
B. bubo	-1.26 0.225				0.15 0.882[s]	2.42 **0.026**	0.349	4.94	3,19	**0.011**
G. passerinum	-2.00 0.060				1.21 0.239[a]		0.128	2.61	2,20	0.099
S. aluco				-2.65 **0.016**	-0.41 0.683[a]	2.62 **0.017**	0.390	5.68	3,19	**0.006**
S. uralensis					3.39 0.003[a]		0.323	11.51	1,21	**0.003**
Nomadic										
F. tinnunculus				-1.79 0.090	-1.94 0.067[s]	1.54 0.140	0.083	1.66	3,19	0.209
A. otus	-2.46 **0.024**				0.38 0.706[s]	3.31 **0.004**	0.440	6.77	3,19	**0.003**
A. funereus	-2.74 **0.013**				0.45 0.661[s]	4.26 **0.000**	0.615	12.72	3,19	**0.000**
Migratory										
B. buteo	-1.32 0.201				-0.34 0.736[s]	2.06 0.054	0.155	2.34	3,19	0.106
C. aeruginosus		-3.24 0.004			2.05 0.0533[a]		0.336	6.58	2,20	**0.006**
Nests										
Sedentary										
B. bubo	-4.06 0.001				0.54 0.593[s]	6.12 **0.000**	0.767	25.08	3,19	**0.000**
S. aluco					3.13 **0.005**[a]		0.286	9.81	1,21	**0.005**
S. uralensis		-2.09 0.050			4.06 **0.000**[a]		0.434	9.42	2,20	**0.001**
Nomadic										
A. funereus	-2.60 **0.017**				-0.04 0.973s	4.25 0.000	0.601	12.06	3,19	0.000
Migratory										
B. buteo	-1.23 0.234				0.10 0.924s	2.09 **0.050**	0.231	3.20	3,19	**0.047**

Large-scale Climatic Phenomena and Local Weather Conditions

Winter conditions may have various kinds of effects on vole-eating birds of prey. The wintering success primarily depends on the availability of food (e.g., Newton 1998). In severe winters the food supply may be limited due to various factors that depend on weather conditions, in particular due to frost and snow cover. Additionally, there may be considerable annual fluctuations in the real amount of food.

Large-scale climatic indices such as those of the North Atlantic Oscillation are associated with population dynamics variation in demographic rates and values of phenotypic traits in many species (Hallett *et al.* 2004). Paradoxically, these large-scale indices can seem to be better predictors of ecological processes than local climate (Forchhammer & Post 2004, Hallett *et al.* 2004). For example, it has been shown, using detailed data from a population, that high rainfall, high winds, or low temperatures at any time during a 3-month period can cause mortality either immediately or lagged by a few days in sheep (Hallett *et al.* 2004). Most measures of local climate used by ecologists fail to capture such complex associations between weather and ecological process, and this may help to explain why large-scale, seasonal indices of climate spanning several months can outperform local climatic factors. Furthermore, the timing of bad weather within a period of mortality can have an important modifying influence on intraspecific competition for food, revealing an interaction between climate and density dependence that the use of large-scale climatic indices or inappropriate local weather variables might obscure (cf. also frost seesaw effect, Solonen 2004).

Conclusion

The significant relationships that emerged are interesting, in particular, because they are revealed by such broad and unstandardized data as mine (see Saurola 2008), and they call for and show the direction to more detailed studies. NAO indicates combined effects of various weather factors while local measurements of single weather variables demonstrate more specific causal chains. Similar associations with environmental variables in different species suggest real causal relationships. From the independent variables examined, I don't know the role of relationships with possibly interacting unknown factors. As shown also in this study, besides the effects of single factors, those of interactions of different variables may be significant.

Besides local factors, large-scale climatic phenomena and population responses of larger geographical areas can be expected to reflect in the dynamics of even small local populations of small mammals. The prevalence of the patterns expressed by small local populations is supposed to be verified by the similarities in the dynamics of small mammals and their avian predators.

My results suggest that the effects of global warming on various vole-eating birds of prey at high latitudes were both positive and negative, in particular due to mild winters. This would lead to changes in local populations and distribution ranges of species. Due to their flexible moving habits, nomadic species might be less seriously affected than site-tenacious ones that are more dependent on local resources, such as nest sites. From a local point of view and during a short period of time, however, the impact seemed to be more pronounced on nomadic species due to their sudden and drastic shifts.

ACKNOWLEDGMENTS

I express my thanks to those bird ringers and ornithologists who have participated in the monitoring program of birds of prey in Uusimaa in 1986–2008. Pentti Ahola (Kimpari Bird Projects), as well as Asko Kaikusalo and Otso Huitu (Finnish Forest Research Institute) provided the vole trapping data, and Martti Heikinheimo (Finnish Meteorological Institute) put the local weather data at my disposal. Pertti Saurola helped by rendering the figures.

REFERENCES

Aars, J. & Ims, R. A. (2002). Intrinsic and climatic determinants of population demography: the winter dynamics of tundra voles. *Ecology, 83*, 3449–3456.

Burnham, K. & Anderson, D. (2002). *Model selection and multimodel inference: a practical information-theoretic approach* (2nd ed.). New York, NY: Springer.

Cramp, S. (ed) (1985). *Handbook of the birds of Europe, the Middle East and North Africa. The birds of the Western Palearctic. Vol. IV.* Oxford, UK: Oxford University Press.

Cramp, S. & Simmons, K. E. L. (Eds.) (1980). *Handbook of the birds of Europe, the Middle East and North Africa. The birds of the Western Palearctic. Vol. II.* Oxford, UK: Oxford University Press.

Christensen, P. & Hörnfeldt, B. (2003). Long-term decline of vole populations in northern Sweden: a test of the destructive sampling hypothesis. *Journal of Mammalogy 84*, 1292–1299.

Drebs, A. Nordlund, A. Karlsson, P. Helminen, J. & Rissanen, P. (2002). *Climatological statistics of Finland 1971–2000.* Helsinki, Finland: Finnish Meteorological Institute.

Forchhammer, M. C. & Post, E. (2004). Using large-scale climate indices in climate change ecology studies. *Population Ecology, 46*, 1–12.

Hallett, T. B. Coulson, T. Pilkington, J. G. Clutton-Brock, T. H. Pemberton, J. M. & Grenfell, B. T. (2004). Why large-scale climate indices seem to predict ecological processes better than local weather. *Nature, 430*, 71-75.

Hansen, T. F. Stenseth, N. C. & Henttonen, H. (1999). Multiannual vole cycles and population regulation during long winters: an analysis of seasonal density dependence. *American Naturalist, 154*, 129-139.

Hanski, I. Hansson, L. & Henttonen, H. (1991). Specialist predators, generalist predators, and the microtine rodent cycle. *Journal of Animal Ecology, 60*, 353-367.

Hanski, I. Henttonen, H. Korpimäki, E. Oksanen, L. & Turchin, P. (2001). Small-rodent dynamics and predation. *Ecology, 82,* 1505-1520.

Hanski, I. & Henttonen, H. (1996). Predation on competing rodent species: A simple explanation of complex patterns. *Journal of Animal Ecology, 65,* 220-232.

Hansson, L. (1999). Intraspecific variation in dynamics: small rodents between food and predation in changing landscapes. *Oikos, 85*, 159–169.

Hansson, L. & Henttonen, H. (1985). Gradients in density variations of small rodents: the importance of latitude and snow cover. *Oecologia, 67,* 394–402.

Hansson, L. Jędrzejewska, B. & Jędrzejewski, W. (2000). Regional differences in dynamics of bank vole populations in Europe. *Polish Journal of Ecology, 48* Suppl., 163–177.

Henttonen, H. (2000). Long-term dynamics of the bank vole *Clethrionomys glareolus* at Pallasjärvi, northern Finnish taiga. *Polish Journal of Ecology, 48* Suppl., 87-96.

Henttonen, H. Oksanen, T. Jortikka, H. & Haukisalmi, V. (1987). How much do weasels shape microtine cycles in northern Fennoscandian taiga? *Oikos, 50,* 353–365.

Hildén, O. (1989). The effects of severe winters on bird fauna of Finland. *Memoranda Societatis pro Fauna et Flora Fennica, 65,* 59–66.

Hipkiss, T. Stefansson, O. & Hörnfeldt, B. (2008). Effect of cyclic and declining food supply on great grey owls in boreal Sweden. *Canadian Journal of Zoology, 86,* 1426–1431.

Hughes, L. (2000). Biological consequences of global warming: is the signal already? *Trends in Ecology and Evolution, 15,* 56-61.

Huitu, O. Koivula, M. Korpimäki, E. Klemola, T. & Norrdahl, K. (2003). Winter food supply limits growth of northern vole populations in the absence of predation. *Ecology, 84,* 2108-2118.

Hämet-Ahti, L. (1981). The boreal zone and its biotic subdivision. *Fennia, 159,* 69–75.

Hörnfeldt, B. (2004). Long-term decline in numbers of cyclic voles in boreal Sweden: analysis and presentation of hypotheses. *Oikos, 107,* 376–392.

Hörnfeldt, B. Hipkiss, T. & Eklund, U. (2005). Fading out of vole and predator cycles? *Proceedings of the Royal Society of London B, 272,* 2045–2049.

Ims, R. A. & Andreassen, H. P. (2000). Spatial synchronization of vole population dynamics by predatory birds. *Nature, 408,* 194–196.

Ims, R. A. Henden, J. A. & Killengreen, S. T. (2008). Collapsing population cycles. *Trends in Ecology and Evolution, 23,* 79-86.

Jędrzejewski, W. & Jędrzejewska, B. (1993). Predation on rodents in Białowieża Primeval Forest, Poland. *Ecography, 16,* 47-64.

Kareiva, P. M. Kingsolver, J. G. & Huey, R. B. (1993). *Biotic interactions and global chance.* Sunderland, UK: Sinauer.

Korpimäki, E. (1985). Rapid tracking of microtine populations by their avian predators: possible evidence for stabilizing predation. *Oikos, 45,* 281–284.

Korpimäki, E. & Norrdahl, K. (1989). Predation of Tengmalm's owls: numerical responses, functional responses and dampening impact on population fluctuations of microtines. *Oikos, 54,* 154–164.

Korpimäki, E. Norrdahl, K. Klemola, T. Pettersen, T. & Stenseth, N. C. (2002). Dynamic effects of predators on cyclic voles: field experimentation and model extrapolation. *Proceedings of the Royal Society of London B, 269,* 991-997.

Korslund, L. & Steen, H. (2006). Small rodent winter survival: snow conditions limit access to food resources. *Journal of Animal Ecology, 75,* 156–166.

Krebs, C. J. & Myers, J. H. (1974). Population cycles in small mammals. *Advances in Ecological Research, 8,* 268–400.

Laaksonen, T. Korpimäki, E. & Hakkarainen, H. (2002). Interactive effects of parental age and environmental variation on the breeding performance of Tengmalm´s owls. *Journal of Animal Ecology, 71,* 23-31.

Lindström, E. & Hörnfeldt, B. (1994). Vole cycles, snow depth and fox predation. *Oikos, 70,* 156-160.

Linkola, P. & Myllymäki, A. (1969). Der Einfluss der Kleinsäugerfluktuationen auf das Brüten einiger kleinsäugerfressender Vögel im südlichen Häme, Mittelfinnland 1952–1966. *Ornis Fennica, 46,* 45–78.

McCarty, J. P. (2001). Ecological consequences of recent climate change. *Conservation Biology, 15,* 320-331.

Merritt, J. F. (1985). Influence of snowcover on survival of *Clethrionomys gapperi* inhabiting the Appalachian and Rocky Mountains of North America. *Acta Zoologica Fennica, 173,* 73-74.

Merritt, J. F. Lima, M. & Bozinovic, F. (2001). Seasonal regulation in fluctuating small mammal populations: feedback structure and climate. *Oikos, 94,* 505-514.

Mikkola, H. (1983). *Owls of Europe.* Calton, UK: Poyser.

Møller, A. P. Fiedler, W. & Caswell, H. (2004). *Birds and climate change. Advances in Ecological Research Vol 35.* London, UK: Elsevier.

Newton, I. (1998). *Population limitation in birds.* San Diego, CA: Academic Press.

Nielsen, J. T. & Møller, A. P. (2006). Effects of food abundance, density and climate change on reproduction in the sparrowhawk *Accipiter nisus. Oecologia, 149,* 505-518.

Norrdahl, K. (1995). Population cycles in northern small mammals. *Biological Reviews, 70,* 621–637.

Nybo, J. O. & Sonerud, G. A. (1990). Seasonal changes in diet of hawk owls *Surnia ulula:* importance of snow cover. *Ornis Fennica, 67,* 45-51.

Pucek, Z. Jędrzejewski, W. Jędrzejewska, B. & Pucek, M. (1993). Rodent population dynamics in a primeval deciduous forest (Białowieża National Park) in relation to weather, seed crop, and predation. *Acta Theriologica, 38,* 199–232.

R Development Core Team (2008). *R: A language and environment for statistical computing.* Vienna, Austria: R Foundation for Statistical Computing. ISBN 3-900051-07-0. URL http://www.R-project.org

Rueness, E. K. Stenseth, N. C. O'Donoghue, M. Boutin, S. Ellegren, H. & Jakobsen, K. S. (2003). Ecological and genetic spatial structuring in the Canadian lynx. *Nature, 425,* 69-72.

Saurola, P. (2008). Monitoring birds of prey in Finland: a summary of methods, trends, and statistical power. *Ambio, 37,* 413-419.

Schwartz, M. D. Ahas, R. & Aasa, A. (2006). Onset of spring starting earlier across the Northern Hemisphere. *Global Change Biology, 12,* 343-351.

Solonen, T. (2004). Are vole-eating owls affected by mild winters in southern Finland? *Ornis Fennica, 81,* 65–74.

Solonen, T. (2005). Breeding of Finnish birds of prey in relation to variable winter food and weather conditions. *Memoranda Societatis pro Fauna et Flora Fennica, 81,* 19–31

Solonen, T. (2006). Overwinter population change of small mammals in southern Finland. *Annales Zoologici Fennici, 43,* 295–302.

Solonen, T. (2008). Large-scale climatic phenomena and timing of breeding in a local population of the Osprey *Pandion haliaetus* in southern Finland. *Journal of Ornithology, 149,* 229–235.

Solonen, T. & Ahola, P. (2010). Intrinsic and extrinsic factors in the dynamics of local small-mammal populations. *Canadian Journal of Zoology, 88(2),* 178-185.

Sonerud, G. A. (1986). Effect of snow cover on seasonal changes in diet, habitat, and regional distribution of raptors that prey on small mammals in boreal zones of Fennoscandia. *Holarctic Ecology*, *9*, 33-47.

Steen, H. Ims, R. A. & Sonerud, G. A. (1996). Spatial and temporal patterns of small-rodent population dynamics at a regional scale. *Ecology*, *77*, 2365–2372.

Stenseth, N. C. (1999). Population cycles in voles and lemmings: density dependence and phase dependence in a stochastic world. *Oikos*, *87*, 427–461.

Stenseth, N. C. Mysterud, A. Ottersen, G. Hurrell, J. W. Chan, K. S. & Lima, M. (2002). Ecological effects of climate fluctuations. *Science*, *297*, 1292–1296.

Strann, K. B. Yoccoz, N. G. & Ims, R. A. (2002). Is the heart of the Fennoscandian rodent cycle still beating? A 14-year study of small mammals and Tengmalm's owls in northern Norway. *Ecography, 25*, 81–87.

Sundell, J. Huitu, O. Henttonen, H. Kaikusalo, A. Korpimäki, E. Pietiäinen, H. Saurola, P. & Hanski, I. (2004). Large-scale spatial dynamics of vole populations in Finland revealed by the breeding success of vole-eating avian predators. *Journal of Animal Ecology*, *73*, 167–178.

Venables, W. N. Smith, D. M. & the R Development Core Team (2008). *An introduction to R.* Version 2.8.0. The R Project for Statistical Computing. Available at: http://www.r-project.org

Visser, M. E. van Noordwijk, A. J. Tinbergen, J. M. & Lessells, C. M. (1998). Warmer springs lead to mistimed reproduction in great tits (*Parus major*). *Proceedings of the Royal Society of London B*, *265*, 1867-1870.

Vitousek, P. M. (1994). Beyond global warming: ecology and global change. *Ecology, 75,* 1861–1976.

Walther, G. R. Post, E. Convey, P. Menzel, A. Parmesan, C. Beebee, T. J. C., Fromentin, J. M. Hoegh-Guldberg, O. & Bairlein, F. (2002). Ecological responses to recent climate change. *Nature, 416*, 389-395.

Watkinson, A. R. Gill, J. A. & Hulme, M. (2004). Flying in the face of climate change: a review of climate change, past, present and future. *Ibis, 146* Suppl. 1, 4–10.

Zachos, J. Pagani, M. Sloan, L. Thomas, E. & Billups, K. (2001). Trends, rhythms and aberrations in global climate 65 Ma to present. *Science, 292*, 686–693.

In: Trends in Ornithology Research
Editors: P. K. Ulrich and J. H. Willett, pp. 121-136

ISBN: 978-1-60876-454-9
© 2010 Nova Science Publishers, Inc.

Chapter 4

INTEGRATING INDIGENOUS KNOWLEDGE OF BIRDS INTO CONSERVATION PLANNING IN NEW GUINEA

William H. Thomas
The New Jersey School of Conservation
Montclair State University, USA

ABSTRACT

It has been difficult to integrate indigenous knowledge into conservation planning. Although indigenous naturalists have accumulated generations of observations concerning their environments, stereotypes concerning their relationship to nature have frustrated attempts to involve indigenous societies in conservation. However, unencumbered by western philosophy, indigenous naturalists have been developing a dynamic view of nature that incorporates connectedness, disturbance and recovery as a normal course of events in the natural world. This non-linear view of nature has only recently emerged as scientific consensus. In this article, I argue that communication between conservationists and indigenous people can be facilitated by using indigenous knowledge of birds to identify the impacts of tradition on biodiversity. Birds are a commonly acknowledged indicator of biodiversity. Because indigenous people have a long-range perspective on the effects of human activity on avian diversity, they can provide a perspective vital to conservation planning. Drawing on ethno-ecological fieldwork with the Hewa of Papua New Guinea, this paper presents an indigenous perspective on the effects of traditional activities on birds. The Hewa describe their traditions as shaping the environment by creating a mosaic of habitats of varying diversity. I argue that the while the current lifestyle of the Hewa may not necessarily be a template for future sustainability, the Hewa view of the natural world provides insights into the potential of indigenous people to conserve their resources.

INTRODUCTION

In 1989 the Coordinating Body of Indigenous Organizations of the Amazon Basin (COICA) issued an appeal to conservation and development organizations asking them to build on the accumulated knowledge of traditional Amazonian societies to guaranteei the future of the Amazon Basin (COICA 1989). Their plea coincided with a surge in the interest in indigenous knowledge (IK) that some had hoped would become a break through for applied anthropology (Sillitoe 1998). When it was issued, the initial reaction was enthusiastic (Chapin 2004). However in the intervening years enthusiasm for partnerships with indigenous communities has waned (Chapin 2004). I hope to reignite the enthusiasm for creating such partnerships by focusing on indigenous knowledge of birds to re-frame the discussion of the relationship between tradition and biodiversity. To do so, I will rely on my research with the Hewa people of Papua New Guinea. Papua New Guinea is the political entity/nation occupying the eastern half of the geographic entity known as the island of New Guinea. The Hewa live on the northern slopes of New Guinea's Central Range around the headwaters of the Strickland River. Their knowledge of these forests has lead me to believe that by focusing on the effect of human activity on birds, we can establish some common ground for the conservation of their lands and a template for others to follow.

The homelands of the remaining indigenous societies contain much of the planet's remaining biological heritage (Robles-Gil 2002). This combined with their reverence for nature, has led many to believe that they are natural conservationists. Indeed, this has been the rationalization for combining sustainable development with conservation, i.e. traditional land management practices have produced the biodiversity we covet. Why not capitalize on these sustainable land management techniques to conserve this biodiversity (Posey 1985)? The compatibility of traditional with biological diversity opened the possibility of an alliance between conservationists and indigenous people (Nabhan 2001). Those advocating this alliance had essentially assumed that indigenous societies had learned to minimize their impact on the land (Smith 1985). This assumption spawned an interest in understanding traditional management systems (Barrett et. al. 2001) ; collecting indigenous knowledge (IK) (Folke 2004; Ludwig etal, 2001); and potentially using traditional practices as templates for biodiversity conservation (Posey 1988).

While most conservationists agree on the need for local participation, this is a question of intent (Parker 1992; Has the conservation of biodiversity by traditional societies been their intention or a side-effect of small-scale human activity on vast landscapes? It quickly became apparent that the importance of local involvement in conservation may have blinded conservationists to the realities of traditional life (Sillitoe 1996). In general, indigenous societies seem to be incapable of conserving game (Redford 1991). Traditional forest management practices may not be compatible with modern conservation (Posey and Balée 1989).

This has caused a backlash among parties who can usually be counted on as advocates for indigenous people. For example, some have pointed to the discrepancy between the western perception of a traditional conservation ethic and reality (Salim et.al.2001). Others have questioned the usefulness of IK in the face of global change (Terborgh 2001; du Toit et al, 2004). Advocates for this alliance may also be laboring under erroneous assumptions concerning the practicality of local participation in the face of national laws that negate local input in the conservation process (Pierce and Wadley 2001). In general, our inability to reconcile traditional life with conservation continues to make the inclusion of indigenous people in the conservation process problematic (Chatty and Colchester 2002). This has prompted conservationist to call for the reconsideration of conservation partnerships with indigenous societies (Soule 2000; Terborgh 2001).

THE RELATIONSHIP BETWEEN BIODIVERSITY AND TRADITIONAL LIFE

I believe that traditional societies can act as stewards of their biological inheritance, even under the stress induced by globalization. It bears repeating here that prior to beginning this experiment of engaging traditional landowners in conservation, no consensus had emerged concerning the relationship between traditional life and biological diversity (West and Brechin 1991). I propose that in order to reverse the current trends, we will first need to deconstruct this latest version of the noble savage. To do so, we should begin by revisiting our most basic assumption concerning nature.

It has been assumed that the most biologically diverse ecosystems were the most stable (Reice 1994). This assumption fit nicely with the traditional western view of the universe as a collection of parts, connected into a machine-like system that has been designed to be predictable, balanced and ultimately controllable by humans (Goerner 1994). By extension, a balanced ecosystem is a stable system in terms of species abundance and composition (Reice 1994). From this mechanistic perspective, the conservation of tropical forests requires that we develop a plan with a complete inventory of the species in the system. This plan should also detail how each species is connected to one another. Once this 'blueprint' for conservation is developed, our goal as conservationist will be to maintain this system of internal relationships – i.e. the balance or equilibrium of the system. So much of the science needed to conserve our natural world consists of understanding these parts, that it seems that we have been developing such blueprints. Yet success in the form of a final product or blueprint ha seluded us. Natural systems have proven to be extremely complex and difficult to define. Ecosystems are rarely, if ever, in a state of equilibrium. After centuries of cataloguing nature's parts, the world continues to be a frustrating machine to keep in balance.

Frustration has led researchers to question their initial premise -- Is the "natural" state of an ecosystem in fact a state of equilibrium? The assumption that ecosystems tend toward equilibrium is so pervasive that this question borders on heresy. However, the difficulty in defining extremely complex systems has led ecologists to abandon the equilibrium model and instead concentrate on the dynamic components of an ecosystem (Pickett et al, 1991). In this dynamic paradigm, the balance of nature concept is described as non-scientific (Pickett and White 1985). Unfortunately, the shift has gone unrecognized by many anthropologists (Smith 1984:3). Authors continue to portray traditional societies as "in balance" or describe a practice as "adaptive." Yet, this use of terms drawn from ecology and evolutionary biology is often outmoded (Hames 1991).

Current research has focused on the role of nonequilibrium factors, commonly referred to as disturbance, in the enhancement of biodiversity (Reice 1994:924). Ecologists define a disturbance as any 'relatively discrete event that disrupts a population, community or ecosystem and changes resources available' (Pickett and White 1985). Disturbance is unpredictable, nonselective and can produce effects that will vary from minutes to centuries in duration (Pickett et al, 1991). Not to be confused with predation (which is 'intrinsic to the life of the prey species'-- prey adapt to it), disturbance is unpredictable and nonselective (Reice 1994:428). It can come in any size, at any time and produce effects that will vary from minutes to centuries in duration. While we typically think of disturbance as phenomena like storms that originate outside of an ecosystem, ecologists have discovered rich, dynamic and unpredictable behavior arising from the internal dynamics of laboratory populations without an external source of disturbance (May 1989:37). Disturbance creates the patchiness that characterizes many environments and this patchiness translates into niches that present opportunities for colonization by new species (May 1989). To the new non-equilibrium school, recovery from disturbance - not equilibrium - is the normal state of affairs in any ecosystem (Reice 1994:427).

The eventual structure of any ecological community is determined by its response to continual disturbance. At either extreme of the disturbance continuum, environments that are either undisturbed or constantly wracked by severe disturbances will eventually be dominated by a few species that have evolved to take advantage of these circumstances (Terborgh 1992:99). As long as disturbances occur frequently enough to prevent the competitive exclusion of poorer competitors, but do not devastate the ecosystem, they serve to enhance biological diversity. Species adapted to the many niches continually remade by disturbance can continue to survive along with more efficient species (Reice 1994:428). By creating a mosaic of environments that prevents the extinction of competing species, disturbance serves to promote a high degree of species richness (Connell 1978). Essentially the linkage between diversity and disturbance can be reduced to this: In terms of its ability to generate biodiversity, disturbance is a scale related phenomena. It enriches two measures of diversity by creating more habitats (gamma diversity) and containing more organisms (greater alpha diversity) than an unaltered landscape. Too much or too little disturbance produces environments that are not as diverse as those that are continually subjected to minor disturbances (Terborgh 1992).

The shift from balance to disturbance by ecologists has gone unrecognized by those who continue to portray traditional societies as 'in balance' with their surroundings (Hames 1991). The underlying assumption of much of the research into the relationship between traditional societies and their environment has been that these non-western societies have learned to minimize their impact, i.e. not disturb the balance of nature. However, not only does the research indicate that ecosystems are rarely, if ever, in a state of equilibrium; it also seems that greater species diversity is found in systems that experience disturbance. Therefore there is no sense in searching for clues to a group of human's with the ability to maintain the natural balance ecosystems that have no inherent tendency toward balance. Instead traditional activities should be examined as possible sources of disturbance.

By factoring disturbance into the relationship between tradition and biodiversity, researchers have moved beyond the stereotypes of the 'noble savage' and are beginning to unravel the archaeological evidence of humanity's role in both historic and prehistoric extinctions (Diamond 1986; Denevan 1992). The continued disappearance of wild lands coupled with the coexistence of traditional cultures with biological diversity, often referred to as biocultural diversity (Maffi 2001), has forced conservationists to reconsider our notions concerning the nature of wilderness (Mittemeyer et.al. 2003). Nowhere has this paradigm shift had more impact than in Amazonia. Here historical ecologists have begun to paint a more nuanced picture of the relationship of traditional societies to biological diversity. Today Amazonian forests are no longer characterized as pristine. Instead, the Amazonian landscape has been described as a tribute to the 'immense transformative power of prehistoric humans' (Graham 1998). The prehistoric Amazon basin once supported larger, sedentary and more complex societies (Roosevelt 1980). Extant Amazonian societies are more mobile than their historical predecessors, and in terms of their ecological impact, resemble hunter-gatherers

(Roosevelt 1989:30). These societies are now seen as master manipulators of their forests, intensely managing their lands produce a mosaic of habitats (Posey 1985; Johnson 1989; Balée 1994). They create and then manage the various vegetative zones using techniques that go beyond clearing and burning (Balée and Gely 1989; Anderson and Posey 1989). While the biological diversity of any plot may initially plummet as it the land is cleared for gardens, this decline in diversity is not necessarily permanent and in many cases will eventually be part of a mosaic with greater diversity than the original landscape (Balée1998). Although the forests of Amazonia are still considered one of the Earth's finest wilderness tracts, for many analysts these forests can best be understood as an old fallow rather than pristine forest (Balée1995). To the extent that practices like gardening are small-scale environmental disturbances that traditionally have created a biologically diverse mosaic, it has important implications for the ability of local people to respond to changes demanded by the boundaries etc. that will accompany a conservation agreement.

THE HEWA

In an attempt to understand the relationship between indigenous people and their environment, I have been recording IK of the Hewa of Papua New Guinea concerning the impact of human activity on birds. The Hewa are a traditional society of swidden horticulturalists whose homeland is one of the island of New Guinea's most important wilderness areas. Since conservation programs in the nation of Papua New Guinea must be generated by the local people, the Hewan understanding of the relationship between tradition and avian/biological diversity will be a vital link in the conservation of this area.

By using the Hewa IK of birds, I hope to describe the relationship between traditional activities and biological diversity in a manner that will be intelligible to both the Hewa and conservationists. In this process, I felt that we would need to take ethnoecologist Gary Nabhan's advice and develop an ethno database that would go beyond species inventories (Nabhan 2001). All ethnobiological studies have to deal with the cultural gaps created by New Guinea vernacular and the genus species system of western science (Berkes and Folke 1998; Diamond and Bishop 1999; Sillitoe 2000). However birds are an accepted indicator of ecosystem diversity and health (Azevedo-Ramos, De Carvalho and Nasi 2002). Since birds are also the best known organisms in New Guinea, I believe that the following approach allows both the Hewa and conservationist to establish a common ground concerning the impact of human activity on avian diversity. It will assist both parties in the evaluation of their capacity for co-management and overcome major obstacles in communication (Drew 2005). This approach should also enable us to avoid some of the more contentious issues surrounding IK (Ellen and Harris 2000).

New Guinea is one of the world's most significant centers of biodiversity and contains large tracts of intact forest (Meyers et al, 2000).). Since it has retained 75% of its' primary vegetation, the island of New Guinea continues to be described by conservationists as a "good news area" (Meyers et al, 2000). The Hewa live in one of New Guinea's most important wilderness areas, the headwaters of the Strickland River in the Central Range (142 30'E, 5 10' S; elevation 500-3000 meters). They number fewer than 2,000 people and inhabit roughly ca. 65,000 hectares of hilly and sub montane forest in the uppermost Strickland River. Their region is located on the eastern verges of the 'Great Rivers Headwaters,' a rain-soaked upland zone in the center of New Guinea that recent analyses have identified as the richest in biodiversity in this island. (Beehler 1993). This Headwaters region is where the four great river systems of New Guinea converge (Sepik, Fly, Digul, Idenburg). The Strickland is the major tributary of the Fly and the Hewa inhabit the forests where the Strickland meets the torrential Laigaip River. The forests in this region are extensive and the land is dominated by a mosaic of primary and secondary growth forest. While there have been no previous studies of the forests in the Hewa territory, the area surrounding headwaters of the Strickland River has been identified as a 'major terrestrial unknown,' but has no formal conservation status (Swartzendruber 1993).

Societies in New Guinea, like their Amazonian counterparts, have been described as developing traditions that enable them to coexist with biological diversity (Sillitoe 1996). However in seventeen years of fieldwork, I have never heard the Hewa use the term 'balance' to describe their relationship to the land. Instead the Hewa describe their traditional activities as creating a mosaic of garden *Agwe*, grassland *Poghali*, old garden *Agwe Teli*, old garden 'true' *Agwe Teli Popi* and primary forest *Nomakale* – each with a set of pollinators and seed dispersal agents that are impacted by the Hewa cutting the forest to establish and maintain gardens. The microclimate associated with altitude and terrain effectively confines Hewa horticulture between the altitudes of 500m at the riverbank and the base of the mountain wall at 1500m., with the majority of these gardens below 1000m. The Hewa raise their gardens, relying primarily on sweet potato (*Ipomoea batatas*), yams (*Dioscorea sp.*), banana (*Musa sp.*) and to a lesser degree cassava (*Manihot esculenta*) and pumpkin (*Cucurbita maxima*) as food crops. Scattered throughout the area are several species of Pandanus and *Pangium edule* trees that the Hewa claim individually. The seasonal ripening of these trees, as well as gathering other wild foods and hunting, provides the Hewa with some sustenance. However gardens are the primary source of food. Each year the typical household clears and plants an average of four $100m^2$ gardens. Like many New Guineans, the Hewa re-use their gardens. In order to use as much of the fence surrounding an old garden as possible, the Hewa cut new gardens adjacent to previous ones, thereby creating a chain of old and new gardens. The established gardens seldom lie fallow for more than twenty five years, at which time their secondary forest cover is cut, burned and cleared and a new garden planted. The result is a mosaic on the surrounding hillsides comprised primarily of primary forest interspersed with small plots of land in the garden/fallow cycle. This mosaic of new gardens, grasslands, succession and primary forest increases the number of environments and hence one measure of the biodiversity of this territory.

Because New Guinea is east of the Wallace line, the island lacks many of the mammalian agents of seed dispersal found to the west in Indonesia. In order to assess the compatibility of the traditional Hewa lifestyle with biodiversity, I have asked my informants to describe the impact of traditional gardening on New Guinea's primary agents of seed dispersal -- birds. Well known to both local and international naturalists, birds are the key to forest conservation in New Guinea (Schodde 1973).

Through a combination of structured interviews, transects, and station surveys, the Hewa IK concerning the impact of traditional activities on birds was recorded . Working with the field guide, *Birds of New Guinea* (Beehler et al, 1986) each informant was asked to identify the birds to be found in their territory, as well as the altitude and habitat each bird favored. Habitats were broadly defined, again using the indigenous categories for garden *Agwe*, grassland *Poghali*, old garden *Agwe Teli*, old garden 'true' *Agwe Teli Popi* and primary forest

Nomakale. The old garden/old garden true distinction described their perception of the differences between the bird life to be found in secondary forest growth that was younger than twenty years (old garden) and secondary growth with more than twenty years (old garden true). The information obtained in interviews was then checked against four months of field surveys (see Beehler et.al. 1987 for an example of the protocols used). This gardening cycle is the most important factor in shaping this environment and has the greatest implications for conservation of these forests

The results of our work are contained in the Table below. It is an attempt to depict the dynamic possibilities of IK for biodiversity conservation and re-engage those interested in partnerships between indigenous landowners and conservationists.

Common Name	Genus	Species	Habitat	Diet	Altitude
Grey Goshawk	Accipiter	novaehollandiae	B	V	N,C
Black-mantled Goshawk	Accipiter	melanochlamys	B	V	N,C
Grey-headed Goshawk	Accipiter	poliocephalus	B	V	N,C
Chinese Goshawk	Accipiter	soloensis	Ng,Lg,B	V	H,N
Feline Owlet-nightjar	Aegotheles	insignis	B	I	A
Mountain Owlet-nightjar	Aegotheles	albertisi	B	I	A
Wattled Brush-turkey	Aepypodius	arfakianus	B	G	N,C
Spotted Catbird	Ailuroedus	melanotis	Lgt,B	A/F	N,C
Azure Kingfisher	Alcedo	azurea	Lg,B	A/V	A
Papuan King-Parrot	Alisterus	chloropterus	B	S	N,C
Bush-Hen	Amaurornis	olovaceus	Lgt,B	I/G	N,C
Macgregor's Bowerbird	Amblyornis	macgregoriae	B	F/I	C
Salvadori's Teal	Anas	waigiuensis	W	W	A
Gurney's Eagle	Aquila	gurneyi	B	V	N,C
Crested Hawk	Aviceda	subcristata	B	V	A
Sulphur-crested Cockatoo	Cacatua	galerita	Lgt,B	S	A
Brush Cuckoo	Cacomantis	variolosus	Ng,Lg,B	I	A
Chestnut-breasted Cuckoo	Cacomantis	castaneiventris	Ng,Lg,B	I	A
Large-tailed Nightjar	Caprimulgus	macrurus	K,Ng	I	H,N
Dwarf Cassowary	Casuarius	bennetti	Lgtf,B	F	A
Greater Black Coucal	Centropus	menbeki	K,Lg,B	A/V	H
Pheasant Coucall	Centropus	phasianinus	Lgt,B	G	A
Dwarf Kingfisher	Ceyx	lepidus	Lg,B	A/V	A
Josephine's Lorikeet	Charmosyna	josefinae	B	N	N,C
Little Red Lorikeet	Charmosyna	pulchella	B	N	A
Red-flanked Lorikeet	Charmosyna	placentis	B	N	A
Pygmy Lorikeet	Charmosyna	wilhelminae	B	N	C
White-eared Bronze-Cuckoo	Chrysococcyx	meyerii	B	I	N,C
Magnificent Bird of Paradise	Cicinnurus	magnificus	Lgt,B	A/F	A
King Bird of Paradise	Cicinnurus	regius	Lgt,B	A/F	A
Shovel-billed Kingfisher	Clytoceyx	rex	B	A/V	A
Glossy Swiftlet	Collocalia	esculenta	Lg,B	I	A
Little Shrike-thrush	Colluricincla	megarhyncha	Lgt,B	A	A
Black Cuckoo-shrike	Coracina	melaena	K,Lg,B	A/F	A
Black-bellied Cuckoo-shrike	Coracina	montana	Lgtf,B	A	A
Black-shouldered Cuckoo-shrike	Coracina	morio	Lgtf,B	A	A
Stout-billed Cuckoo-shrike	Coracina	caeruleogrisea	Lgtf,B	A/F	A
Boyer's Cuckoo-shrike	Coracina	boyeri	Lgtf,B	A/F	A
Grey-headed Cuckoo-shrike	Coracina	schisticeps	Lgtf,B	F	A
Hooded Butcherbird	Cracticus	cassicus	K,Lg,B	A/F	H,N

(Continued)

Common Name	Genus	Species	Habitat	Diet	Altitude
Black Butcherbird	Cracticus	quoyi	Lg,B	G	A
Bicoloured Mouse Warbler	Crateroscelis	nigrorufa	Lgt,B	I	A
Rusty Mouse Warbler	Crateroscelis	murina	Lgt,B	I	A
Oriental Cuckoo	Cuculus	saturatus	Lg,B	I	A
Double-eyed Fig-Parrot	Cyclopsitta	diophthalma	Lgf,B	F	A
Rufous-bellied Kookabura	Dacelo	tyro	Lgt,B	A/V	H,N
Plumed Whistling Duck	Dendrocygna	eytoni	W	W	A
Papuan Flowerpecker	Dicaeum	pectorale	Ng,Lg,B	A/F	A
Spangled Drongo	Dicrurus	hottentottus	Ng,Lg,B	I	H,N
Northern Scrub-Robin	Drymodes	superciliaris	B	I	A
Purple-tailed Imperial Pigeon	Ducula	rufigaster	B	F	N,C
Zoe Imperial Pigeon	Ducula	zoeae	Lgtf,B	F	A
Eclectus Parrot	Eclectus	roratus	Lgtf,B	S/F	N,C
White-faced Heron	Egretta	novaehollandiae	W	V	A
Pied Heron	Egretta	picata	W	V	A
Little Egret	Egretta	garzetta	W	V	A
Great Egret	Egretta	alba	W	V	A
Common Koel	Eudynamys	scolopacea	Lgt,B	I/G	A
Wattled Ploughbill	Eulacestoma	nigropectus	B	I	N,C
Dollarbird	Eurystomus	orientalis	Lg,B	I/V	A
Brown Falcon	Falco	berigora	B	V	N,C
White-bibbed Ground-Dove	Gallicolumba	jobiensis	B	F	N,C
Cinnamon Ground-Dove	Gallicolumba	rufigula	Lgtf,B	S	A
Blue-collared Parrot	Geoffroyus	simplex	B	S	N,C
Red-cheeked Parrot	Geoffroyus	geoffroyi	B	S/F	N,C
Fairy Gerygone	Gerygone	palpebrosa	K,Lg	I	A
Green-backed Gerygone	Gerygone	chloronotus	K,Lg	I	A
Yellow-bellied Gerygone	Gerygone	chrysogaster	Lg,B	I	A
Papuan Mountain Pigeon	Gymnophaps	albertisii	B	F	N,C
Mountain Kingfisher	Halcyon	megarhyncha	Lgf,B	A/V	A
Forest Kingfisher	Halcyon	macleayii	Lgf,B	A/V	A
Sacred Kingfisher	Halcyon	sancta	Lgf,B	A/V	A
Whistling Kite	Haliastur	sphenurus	Ng,Lg,B	A/V	A
Brahminy Kite	Haliastur	indus	W,Ng	A/V	A
New Guinea Harpy-Eagle	Harpyopsis	novaeguineae	Lgt,B	V	A
Moustached Tree-swift	Hemiprocne	mystacea	Ng,Lg,B	I	A
Long-tailed Buzzard	Henicopernis	longicauda	B	V	A
New Guinea Bronzewing	Henicophaps	albifrons	Lgt,B	S/F	A
Black-browed Triller	Lalage	atrovirens	K,Lg,B	A/F	A
Papuan Hanging Parrot	Loriculus	aurantiifrons	Lgt,B	N/I	A
Western Black-capped Lory	Lorius	lory	Lg,B	N	A
Black-billed Cuckoo-Dove	Macropygia	nigrirostris	B	F	N,C
Brown Cuckoo-Dove	Macropygia	amboinensis	Lgtf,B	F	A
Emperor Fairy-wren	Malarus	cyanocephalus	Ng,Lg,B	I	H,N
Broad-billed Fairy-wren	Malarus	grayi	Ng,Lg,B	I	H,N
White-shouldered Fairy-wren	Malurus	alboscapulatus	Ng,Lg,B	I	H,N
Trumpet Manucode	Manucodia	keraudrenii	Lg,B	F	A
Crinkle-collared Manucode	Manucodia	chalybata	Lg,B	F	A
New Guinea Flightless Rail	Megacrex	inepta	K	I/G	H,N
Common Scrubfowl	Megapodius	freycinet	B	G	N,C
Doria's Hawk	Megatriorchis	doriae	B	V	A
Black Berrypecker	Melanocharis	nigra	Lgtf,B	F	A
Ornate Melectides	Melectides	torquatus	B	N/A	N,C
Belford's Melectides	Melidectes	belfordi	B	N/A	C
Yellow-browed Melectides	Melidectes	rufocrissalis	B	N/A	C
Long-billed Honeyeater	Melilestes	megarhynchus	K,Lg,B	F/I	A
Scrub White-eared Meliphaga	Meliphaga	albonotata	K,Lg,B	F/I	A
Common Smoky Honeyeater	Melipotes	fumigatus	B	F	N,C
Blue-tailed Bee-eater	Merops	philippinus	Ng,Lg,B	I	H,N

Integrating Indigenous Knowledge of Birds into Conservation Planning ... 129

Yellow-legged Flycatcher	Microeca	griseoceps	Lgtf,B	I	A
Olive Flycatcher	Microeca	flavovirescens	Lgtf,B	I	A
Buff-faced Pygmy-Parrot	Micropsitta	pusio	Lgt,B	L	N,C
Black Kite	Milvus	migrans	Ng,Lg,B	A/V	A
Torrent Flycatcher	Monachella	muellerianna	W	I	A
Spot-winged Monarch	Monarcha	guttula	Lgt,B	I	A
Black-winged Monarch	Monarcha	frater	B	I/A	A
Golden Monarch	Monarcha	chrysomela	Lg,B	I/A	A
Grey Wagtail	Motacilla	cinerea	K,Lg,Ng	I	A
Satin Flycatcher	Myiagra	cyanoleuca	Lgt,B	I	A
Red Myzomela	Myzomela	cruentata	B	N/I	C
Red-throated Myzomela	Myzomela	eques	B	N/I	C
Mountain Red-headed Myzomela	Myzomela	adolphinae	B	N/I	C
Papuan Black Myzomela	Myzomela	nigrita	B	N/I	C
Yellow-bellied Sunbird	Nectarina	jugularis	Lgtf,B	N/A	A
Rufous Owl	Ninox	rufa	B	V	A
Pygmy Honeyeater	Oedistoma	pygmaeum	K,Lg	N/I	A
Dwarf Honeyeater	Oedistoma	iliolophus	K,Lg	N/I	A
Brown Oriole	Oriolus	szalayi	K,Lg,B	F/I	A
Pheasant Pigeon	Otidiphaps	nobis	B	S/F	N,C
Dwarf Whistler	Pachycare	flavogrisea	B	I	N,C
Black-headed Whistler	Pachycephala	monacha	Lg,B	I	A
Rusty Whistler	Pachycephala	hyperythra	Lg,B	I	A
Sclater's Whistlers	Pachycephala	soror	Lgt,B	I	A
Golden-backed Whistler	Pachycephala	aurea	Ng,Lg,B	I	A
White-eyed Robin	Pachycephalopsis	poliosoma	B	I	N,C
Short-tailed Paradigalla	Paradigalla	brevicauda	B	A/F	C
Raggianna Bird of Paradise	Paradisaea	raggiana	Lgf,B	A/F	A
Mountain Peltops	Peltops	montanus	Ng,Lg,B	I	A
Helmeted Friarbird	Philemon	buceroides	K,Lg,B	F/I	A
Island Leaf-Warbler	Phylloscopus	trivirgatus	Lg,B	I	N,C
Crested Pithoui	Pithoui	cristatus	B	A	A
Rusty Pithoui	Pitohui	ferrugineus	K,Lg,B	F/I	A
Hooded Pithoui	Pitohui	dicrous	Lg,B	A/F	A
Blue-breasted Pitta	Pitta	erythrogaster mackloti	Lgtf,B	I/A	A
Hooded pitta	Pitta	sordida	Lgtf,B	I/A	A
Noisy Pitta	Pitta	versicolor	Lgtf,B	I/A	A
Papuan Frogmouth	Podargus	papuensis	Lgf,B	I/V	A
Marbled Frogmouth	Podargus	ocellatus	Lgf,B	I/V	A
Spotless Crake	Porzana	tabuensis	K,Kt	I/G	H
Palm Cockatoo	Probosciger	aterrimus	B	S/V	A
Dusky Lorikeet	Pseudeos	fuscata	Lg,B	N	A
Vulturine Parrot	Psittrichas	fulgidus	B	F	N,C
King of Saxony Bird of Paradise	Pteridophora	alberti	B	F/I	C
White-breasted Fruit-Dove	Ptilinopis	rivoli	B	F	N,C
Pink-spotted Fruit-Dove	Ptilinopus	perlatus	B	F	A
Ornate Fruit-Dove	Ptilinopus	ornatus	B	F	A
Dwarf Fruit-Dove	Ptilinopus	nanus	B	F	A
Beautiful Fruit-Dove	Ptilinopus	pulchellus	Lg,B	F	A
Wompoo Fruit Dove	Ptilinopus	magnificus	Lgtf,B	F	A
Superb Fruit-Dove	Ptilinopus	superbus	Lgtf,B	F	A
SpottedJewel-Babbler	Ptilorrhoa	leucosticta	B	A	A
Chestnut-backed Jewel-babbler	Ptilorrhoa	castanonotus	B	A	A
Blue Jewel-Babbler	Ptilorrhoa	caerulescens	B	A	A
Red-necked Rail	Rallina	tricolor	K,Lg,B	I/G	A
Forbes' Forest-Rail	Rallina	forbesi	K,Lg,B	I/G	A
Great Cuckoo-Dove	Reinwardtoena	reinwardtii	B	F	A
Spotted Berrypecker	Rhamphocharis	crassirostris	Lgt,B	F	A
Willie Wagtail	Rhipidura	leucophrys	K	I	A

130 William H. Thomas

(Continued)					
Northern Fantail	Rhipidura	rufiventris	Lg,B	I	A
Black Fantail	Rhipidura	threnothorax	Lgt,B	I	H,N
Chestnut-bellied Fantail	Rhipidura	hyperthra	Ng,Lg,B	I	A
Rufous-backed Fantail	Rhipidura	rufidorsa	Ng,Lg,B	I	A
Papuan Hornbill	Rhyticeros	plicatus	B	F/G	A
Channel-billed Cuckoo	Scythrops	novaehollandiae	Ng,Lg,B	I/G	H,N
Perplexing Scrub-wren	Sericornis	virgatus	Lgt,B	I	A
Beccari's Scrub-wren	Sericornis	becarii	Lgt,B	I	A
Grey-green Scrub-wren	Sericornis	arfakianus	Lgt,B	I	A
Flame Bowerbird	Sericulus	aureus	B	A/F	C
Wallace's Fairy-wren	Sipodotus	wallacii	Ng,Lg,B	I	H,N
Brown-collared Brush-turkey	Talegalla	jobiensis	B	G	A
Little Paradise-Kingfisher	Tanysiptera	hydrocharis	B	A/V	N,C
Common Paradise-Kingfisher	Tanysiptera	galatea	B	A/V	N,C
Slaty-chinned Longbill	Toxorhamphus	poliopterus	Lgtf,B	N/A	A
White-faced Robin	Tregellasia	leucops	B	I	A
Rainbow Lorikeet	Trichoglossus	haematodus	Lg,B	N	A
Thick-billed Ground-Pigeon	Trugon	terrestris	Lgt,B	S/F	H
Grass Owl	Tyto	capensis	B	V	A
Sooty Owl	Tyto	tenebricosa	B	V	A
Tawny-breasted Honeyeater	Xanthotis	flaviventer	K,Lg,B	G	A
New Guinea White-eye	Zosterops	novaeguinneae	Lgt,B	G	A
Black-fronted White-eye	Zosterops	artifrons	Lgt,B	G	A
Western Mountain White-eye	Zosterops	fuscicapillus	Lgt,B	G	A

TABLE CODES FOR HEWA BIRDS
Altitude Codes:

A = all elevations
C = 1000 m. +
N/C = 800 m. +
Habitat Codes:

B = primary Forest
Lgt = forest 20 yrs. +
Lgtf=fewfoundinforest+20yr
Lg = secondary forest
Ng = new garden
K = grassland
W = water
H = 500-800m.
Diet Codes:

L = lichens
V = vertebrates
I = insects
G = generalists
F = fruit
A = arthropods
S = seeds
N = nectar
W = feeds on waterborne vegetation an creatures

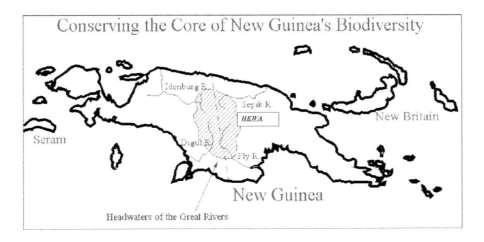

Like western ornithologists, the Hewa associate species with altitude and habitat. Although the western genus and species classifications do not correspond to the Hewa folk taxa (to date the 179 western species correspond to 128 folk taxa). For my purposes, it is more important that my Hewa informants recognize the impact that human disturbance of the primary will forest will have on avian diversity. As expected, the Hewa are keenly aware of the linkage between a birds and habitat. They indicate that some species are associated exclusively with primary forest and that others can make use of forests that the Hewa describe as the oldest secondary forest growth, i.e. forest that has been growing for twenty or more years. Experience has taught the Hewa that cutting the primary forest will eliminate at least thirty-three percent of birds (fifty six species) that can only live in primary forest. The effect of gardening on the habitat preferences of fruit and nectar eating birds is particularly important to biodiversity conservation because, if the scale of habitat modification/disturbance is of sufficient magnitude, the Hewa will compromise the forest's ability to regenerate by limiting the habitats preferred by the agents of regeneration -- fruit and nectar eating birds.

According to the Hewa, frugivores are rare in secondary forest growth that is younger than twenty years. Their gardens create an environment that is hostile to the fruit-doves (*Ptilinopus* sp.) and some species of lorikeets *(Charmosyna* sp.). Both species are vital to forest regeneration. In general, the Hewa report that human disturbance creates environments that are hostile to many species identified exclusively with New Guinea's forests. The vulturine parrot *Psittrichas fulgidus*, pheasant pigeon *Otidiphaps nobis* , blue-collared parrot *Geoffroyus simplex,* wattled brush turkey *Aepypodius arfakianus*, hornbill *Rhyticeros plicatus*, flame bowerbird *Sericulus aureus* and purple-tailed imperial pigeon *Ducula rufigaster* are just a few of the species that the Hewa say will find secondary growth incompatible with their needs.

DISCUSSION

Using IK to produce a western-style scientific product is both theoretically and practically problematic (Bulmer 1969). However by gathering IK on the impacts of human activity, conservationists can obtain information on forest dynamics that is verifiable using site surveys, but would require decades to gather by conventional research methods. My informants have put Hewa land use in a context that illustrates the perils of combining the conservation of their forests with development. Rather than portraying themselves as capable of performing a super human balancing act, the Hewa describe their traditional gardening purely as a source of disturbance on this landscape. At the current level of gardening, the Hewa actually increase the biodiversity of their lands. By felling the forest, they create a mosaic of primary forest, secondary forest, grasslands, gardens and the various phases of succession growth (gamma diversity). They also create habitats for organisms that cannot survive in the primary forest (alpha diversity). They are living example of the tremendous biological diversity that can be produced by a small human population with limited technology continually moving while cutting gardens (Balée 1998). By gardening in the forest, it is possible for the Hewa to create a landscape contains more organisms and more habitats than an unaltered landscape. In this sense, the Hewa are inextricably linked to the biological diversity found in their homeland.

A forest containing the type of small-scale gardening currently practiced by the Hewa is a mosaic of many types of biological communities. The combination of gardens, grasslands, the various stages of forest re-growth and primary forest are more biologically diverse than the climax forest alone. Some species, like humans, are better adapted to take advantage of the succession stages of forest re-growth produced by disturbance. However, gardening creates areas that are lower in biodiversity than the surrounding primary forest, as well as a succession environment that will not be visited by most of the fruit and nectar eating birds this forest depends on for regeneration. Known as beta diversity, this comparative measure of biological diversity can serve as a warning that each stage of forest regeneration will be less diverse than the primary forest. The Hewa are creating habitats each of which is less diverse than the primary forest. While the current mosaic is more diverse than mountains covered exclusively in undisturbed primary forest, the replacement of primary forest by secondary growth will diminish the biodiversity of the Hewa homeland.

The Hewa knowledge of this dynamic provides an important insight into the ability of indigenous man to use the environment without compromising the biological diversity.and has important implications for the ability of local people to conserve biological diversity in the face of changing conditions. The current diversity surrounding most forest dwellers is a by-product of traditional land use by a small, mobile society (Smith & Wishnie 2000). One needs look no further than the highland valleys to the south of the Hewa to find examples of the inability of tradition to curb habitat destruction when faced with novel circumstances. Communitiesliving in the valleys surrounding Lake Kopiago, Mt. Hagen and Wabag, simplified their environments long before the arrival of Europeans. They took advantage of the bounty that accompanied the introduction of the sweet potato in the 16th century to grow in numbers and spread across these valleys. As their gardens increased, they transformed and simplified these once diverse landscapes. Their remaining biological diversity has been relegated to those areas too steep or too cold to be gardened profitably. While these valleys remain culturally interesting, they have become ecologically impoverished.

Yet by using IK to map the effects of human disturbance, indigenous people can earn a seat at the conservation table.. By incorporating disturbance into our models, we can begin to understand how traditions can both promote biodiversity and cause extinctions. I believe that the while societies like the Hewa remain intact and their traditions are compatible with biological diversity, the knowledge exists to develop a land use plan that is truly sustainable.

This will not require vast sums of money or spaceage equipment. Using only traditional knowledge of birds, it is possible to establish the connection between disturbance and biological diversity for other cultures in other areas. In so doing, indigenous landowners and conservation organizations can develop partnerships based on mutual understanding and trust for the sake of planet's remaining 'good news areas.'

ACKNOWLEDGMENTS

The author wishes to express his thanks to both the National Geographic Society, Conservation Internation, The Barrick Corporation, Porgera Joint Venture and the Explorers Club for their support of this research.

REFERENCES

Anderson, A. & Posey, D. (1989). Management of a Tropical Scrub Savanna by the Goroite Kayapo of Brazil, in D. Posey & W. Balée (Eds.) *Resource Management in Amazonia: Indigenous and Folk Strategies. Advances in Economic Botany, vol. 7.* Bronx: New York Botanical Gardens.

Azevedo, C. de Carvalho, O. & Nassi, R. (2002). Animal Indicators, a Tool to Assess Biotic Integrity After Logging Tropical Forests? CIFOR report.

Balée, W. (1994). *Footprints of the Forest: Kaapor ethnobotany—the Historical Ecology of Plant Utilization by an Amazonian People.* Columbia University Press, New York.

Balée, W. (1995). Historical Ecology of Amazonia, in L. Sponsel (Eds.) in *Indigenous Peoples and the Future of Amazonia: An Ecological Anthropology of an Endangered World.* University of Arizona Press, Tucson.

Balée, W. (1998). *Advances in Historical Ecology.* Columbia University Press, New York.

Balée, W. & Gely, A. (1989). Managed Forest Succession in Amazonia: The Kaapo Case, in D. Posey & W. Balée (Eds.) *Resource Management in Amazonia: Indigenous and Folk Strategies. Advances in Economic Botany, vol. 7.* Bronx: New York Botanical Gardens.

Barrett C. B. Brandon, K. Gibson, C. & Gjertsen H. (2001).Conserving tropical biodiversity amid weak institutions. *Bioscience, 51,* 497-502.

Beehler B. (1993). A Biodiversity Analysis for Papua New Guinea With an Assessment for Conservation Needs. Biodiversity Support Program, Washington, D.C.

Beehler, B. Krishna Raju, K. S. R. & Ali, S. (1987). Avian use of Man-disturbed Forest Habitats in the Eastern Ghats, India. *IBIS, no 129,* 197-211.

Beehler B. Pratt, T. & Zimmerman, D. (1986). *Birds of New Guinea,* Princeton: Princeton University Press.

Berkes, F. & Folke, C. (2002). Rediscovery of traditional ecological knowledge as adaptive management, *Ecological Applications, no 10,* 1251-1262.

Bulmer, R. N. H. (1969). *Field Methods in Ethno-Zoology with Special Reference to the New Guinea Highlands.* University of Papua New Guinea.

Chapin, M. (2004). A Challenge to Conservationists, *World Watch magazine,* November/December 2004, 17-31.

Chatty D. & Chochester M. (2002). Introduction, in D. Chatty & M. Colchester (Eds.) *Conservation and Mobile Indigenous Peoples*, Berghahn, New York.

Coordinadora de Organaciones Indigenous de la Cuenca Amazonia /COICA. (1989). Two Agendas for Amazonian Development *Cultural Survival Quarterly, vol 13, no 4,* 75-78.

Connell J. H. (1978). Diversity in tropical rainforests and coral reefs, *Science, no 199,* 1302-1310.

Denevan, W. (1992). The Pristine Myth: the landscape of the Americas in 1492, *Annals of the Association of American Geographers, vol 82, no 3,* 369-385.

Diamond, J. (1986). The Environmentalist Myth, *Nature, no 324,* 19-20.

Diamond, J. & Bishop, K. (1999). Ethno-ornithology of the Ketengban people, Indonesian New Guinea, in D. Medin & S. Attran (Eds.) *Folkbiology*, Bradford Books, New York.

Drew, J. (2005). Use of Traditional Ecological Knowledge in Marine Conservation, *Conservation Biology vol 19, no 4,* 1286-93.

Du Toit, J. T. Walker, B. H. & Campbell, B. M. (2004). Conserving tropical nature: current challenges for ecologists *Trends in Ecology and Evolution, no 19,* 12-17.

Ellen, R. F. & Harris, H. (2000). Introduction, in R. F. Ellen, P. Parkes & A. Bicker (Eds.) *Indigenous Environmental Knowledge and its Transformations*, Harwood Academic Publishers, Amsterdam.

Folke, C. (2004). "Traditional knowledge in socio-ecological systems", *Ecology and Society, vol. 9, no 3,* 7.

Gillett, J. E. (1991). The Health of Women in Papua New Guinea, Papua New Guinea Institute of Medical Research. Monograph *No. 9.*

Goerner, S. J. (1994). *Chaos and The Evolving Ecological Universe* Gordon & Breach Publishing, Langhorne Pennsylvania.

Graham, E. (1998). Metaphor and Metamorphism: some thoughts on environmental Metahistory, in W. Balée (Eds.) *Advances in Historical Ecology*, Columbia University Press, New York.

Hames R. (1991). Wildlife Conservation in Tribal Societies, in M. Oldfield & J. Alcorn (Eds.) *Biodiversity Culture, Conservation and Ecodevelopment*, Westview Press, Boulder Colorado.

Johnson, A. (1989). How the Machinguenga manage resources: conservation or exploitation of nature?, in D. Posey & W. Balée (Eds.) *Resource Management in Amazonia: Indigenous and folk strategies. Advances in Economic Botany vol 7.* New York Botanical Gardens, Bronx, New York.

Ludwig D. Mangel, M. & Haddad, B. (2001). Conservation and Public Policy, *Annual Review of Ecology and Systematics, no 32,* 481-517.

Maffi, L. (2001). *On Biocultural Diversity: linking language, knowledge and the Environment*, Smithsonian Institute Press, Washington D.C.

May, R. (1989). The Chaotic Rhythms of Life, *New Scientist*, 18 November, 37-41.

Meyers, N. Mittermeier, R. Mittermeier, C. da Fonseca, G. & Kent, J. (2000). Biodiversity hotspots for conservation priorities, *Nature, no 403,* 853-58.

Mittermeier, R. Mittermeier, C. Brooks, T. M. Pilgrim, J. D. Konstant, W. R. da Fonseca, G. A. B. & Kormos, C. (2003). Wilderness and biodiversity conservation, *Proceedings National Academy Science, vol 100, no 18,* 10309-10313.

Nabhan, G. (2001). Cultural Perceptions of Ecological Interactions An Endangered Peoples Contribution to the Conservation of Biological & Linguistic Diversity, in L. Maffi (Eds.)

On Biocultural Diversity: linking language, knowledge and the environment, Smithsonian Institute Press, Washington D.C.

Parker, E. (1992). Forest Islands and Kayapo Resource management in Amazonia: A Reappraisal of the Apete, *American Anthropologist, 94*, 406-428.

Pickett, S. & White, P.(1985). Introduction, in Pickett S. T. A., White P. S. (Eds.) *The Ecology of Natural Disturbance and Patch Dynamics*, Academic Press, New York.

Pickett, S. Parker, V. & Fiedler, P. (1991). The New Paradigm in Ecology, in P. Fiedler & S. Jain (Eds.) *Conservation Biology*, Chapman and Hall, New York.

Pierce, C. J. & Wadley, R. L. (2001). "From Participation to Rights and Responsibilities in Forest Management: Workable Methods and Unworkable Assumptions in West Kalimantan, Indonesia," in C. J. Pierce Colfer & Y. Byron (Eds.) *People Managing Forests*: *The Links between Human well-Being and Sustainability*, Resources for the Future: Washington D.C.

Posey, D. A. (1985). Indigenous Management of tropical forest ecosystems: the case of the Kayapó Indians of the Brazilian Amazon, *Agroforestry Systems, no 3,* 139-158.

Posey, D. A. (1988). Native and indigenous guidelines for new Amazonian development strategies: understanding biological diversity through ethnoecology, in J. Hemming (Eds.) *Change in the Amazon*, Manchester University Press, Manchester.

Posey, D. A & Balée, W. (1989). Resource Management in Amazonia: Indigenous and folk strategies, *Advances in Economic Botany, vol. 7*, New York Botanical Gardens, Bronx, New York.

Redford, K. H. (1991). The Ecologically Noble Savage, *Cultural Survival Quarterly vol 13, no 1,* 46-48.

Reice S. R. (1994). Nonequilibrium Determinants of Biological Community Structure, *American Scientist, vol 82, no 5,* 424-435.

Robles G. (2002). *Wilderness*, Toppan, Japan.

Roosevelt, A. (1980). *Parmana: Prehistoric maize and manioc along the Amazon and Orinoco*, Academic Press, New York.

Salim, A. Brocklesby, M. A. Tiani, A. M. Tchikangwa, B. Sardjono, M. A. Porro, R. Woelfel, J. & Pierce Colfer, C. J. (2001). In Search of a Conservation Ethic, in C. J. Pierce Colfer & Y. Byron (Eds.) *People Managing Forests*: *The Links between Human Well-Being and Sustainability*. Resources for the Future: Washington D.C.

Schodde R. (1973). General problems of Faunal Conservation in Relation to the Conservation of Vegetation in New Guinea, in A. B. Costin & R. Groves (Eds.) *Nature Conservation in the Pacific*, ANU Press, Canberra.

Sillitoe, P. (1996). *A Place Against Time: Land and environment in the Papua New Guinea Highlands,* Harwood Academic, Amsterdam.

Sillitoe, P. (1998). The Development of Indigenous Knowledge, *Current Anthropology vol 39, no 2,* 223-252.

Sillitoe, P. (2000). Let Them Eat Cake: Indigenous Knowledge, Science and the Poorest of the Poor *Anthropology Today, vol 16, no 6,* 3-7.

Smith, E. A. (1984). Anthropology, Evolutionary Ecology and the Explanatory Poverty of the Ecosystem Concept, in E. Moran (Eds.) *The Ecosystem Concept in Anthropology*, Westview Press, Boulder Colorado, USA.

Smith, E. A. & Wishnie, M. (2000). Conservation and Subsistence in Small-Scale Societies, *Annual Review of Anthropology, vol 29,* 493-524.

Soule, M. (2000). Does Sustainable Development Help Nature?, *Wild Earth, vol 10, no 4,* 56-63.

Swartzendruber, J. F. (1993). *Conservation Needs Assessment*, USAID. Biodiversity Support Program. Washington D.C.

Terborgh, J. (2001). Why Conservation in the Tropics is Failing, in D. Rothenberg & M. Ulvaeus (Eds.) *The World and the Wild: Expanding Wilderness Conservation Beyond its American Roots*, Tucson, University of Arizona Press.

Terborgh, J. (1992). *Diversity and the Tropical Rainforest*, Scientific American Library New York.

West P. & Brechin S. (1991). Introduction, in P. West & S. Brechin (Eds.) *Resident Peoples and National Parks*, University of Arizona Press, Tucson.

In: Trends in Ornithology Research
Editors: P. K. Ulrich and J. H. Willett, pp. 137-151

ISBN: 978-1-60876-454-9
© 2010 Nova Science Publishers, Inc.

Chapter 5

EUROPEAN BIRD SPECIES HAVE EXPANDED NORTHWARDS DURING 1950-1993 IN RESPONSE TO RECENT CLIMATIC WARMING

Gregorio Moreno-Rueda[*]

Departamento de Biología Animal, Facultad de Ciencias, Universidad de Granada, E-18071, Granada, Spain
and
Konrad Lorenz Institut für Vergleichende Verhaltensforschung, Österreichische Academie der Wissenschaften, Savoyenstraße 1a,
A-1160, Wien, Austria

ABSTRACT

Because temperature affects the distribution limits of many organisms, global warming may provoke an advance of distribution ranges polewards. This work examines whether European birds have advanced their distribution ranges mainly northwards from 1950 to 1993 in response to climatic warming. Bird species were separated into different categories according to their distribution. The findings show that European birds advanced their distribution ranges northward more than southward. Only northernmost species showed the contrary pattern, probably because their northward expansion was constrained by the Arctic Sea. Birds from the central Europe advanced their distribution ranges primarily northward, strongly suggesting that the change in distribution was occasioned by climate warming, as a change due to other causes predicts equal frequency of species advancing southwards and northwards.

---

[*] Corresponding author: E-mail: gmr@ugr.es

INTRODUCTION

Earth temperature has been generally increasing since the 1970s (IPCC 2001, Houghton 2004). Because many factors of an organism's biology are temperature-dependent, organisms have responded to climatic warming with changes in distribution, phenology and/or morphology (Hughes 2000, Walther *et al.* 2002, Parmesan & Yohe 2003, Root *et al.* 2003, Parmesan 2006). Climate change may also provoke the extinction of many species (Thomas *et al.* 2004), but, for each species, this will depend on its response capacity to such change. A shift in distribution in accordance with the climatic changes may be, *a priori*, a quick and operative response to climatic warming in order to reduce the risk of extinction. For this reason, it is predicted that organisms should expand their distribution ranges polewards in response to climate change. There is evidence that this has occurred in different plants and animals (Walther *et al.* 2002, Parmesan & Yohe 2003, Hickling *et al.* 2006, Parmesan 2006). The capacity of colonization of new zones, on the other hand, also depends on aspects other than temperature, such as the presence of certain resources, competitors, or corridors, which can encourage or limit the expansion capacity of organisms (Hill *et al.* 2001).

Birds probably respond quickly to climatic warming by changing their distribution ranges, because they are highly mobile and their distribution limits are frequently determined by temperature (Root 1988). There is evidence that North-American bird species have expanded poleward in response to recent climate change (Hitch & Leberg 2007, La Sorte & Thompson 2007). In Europe, many bird species expanded their distribution ranges northwards between 1950 and 1993 (Burton 1995), and it is predicted that this expansion will continue during the 21st century in response to climatic warming (Huntley *et al.* 2006). However, an expansion northwards may occur as a consequence of an expansion in the distribution ranges of populations occasioned by causes other than climatic change, hindering the attribution of these changes in distribution to global warming (Parmesan 1996). Southern distribution ranges of many European birds are limited by the Mediterranean Sea (see distribution maps, for example, in Cramp 1998); therefore, if these birds are expanding their distribution ranges, they must necessarily advance northwards. Clear evidence that European bird species have responded to climate change by moving northward is restricted mainly to northern regions, such as United Kingdom (Thomas & Lennon 1999, Hickling *et al.* 2006), and Finland (Brommer 2004). Evidence for such response in Mediterranean regions are scarce (Sanz 2002) and reduced to some species (e.g., Carrillo *et al.* 2007).

In this study, I examine whether, during the period 1950-1993, European birds have changed their distribution ranges directionally northwards, as predicted according to the climatic warming. For this, I re-analyse the data in Appendix 1 of Burton (1995) using statistical methods and, to discern between a response to climatic warming and the effect of a change in distribution due to other causes (for example, habitat modification, etc.), I classified European birds into six categories according to their distribution: birds distributed in the south (S), in the south and centre (SC), only in the centre (C), in the centre and north (CN), in the north (N), and in the south, centre, and north of Europe (SCN). If changes in distribution are due to climatic change (Hypothesis 1), I predicted an advance mainly northward of the distribution limits for birds in all categories, although perhaps this advance is constrained for the northernmost species (Table 1). However, if changes in distribution occurred for other causes (Hypothesis 2), I predicted that the distribution of southern species (S and SC) would advance primarily northwards, the distribution of northern species (N and CN) would advance primarily southwards, and the distribution of central species would advance southwards and northwards with similar frequency of species (Table 1) (for similar predictions see, for example, Parmesan *et al.* 1999, Thomas & Lennon 1999, Brommer 2004, Hitch & Leberg 2007).

METHODS

Information on changes in distribution for European birds between 1950 and 1993 was derived from Appendix 1 in Burton (1995). This Appendix shows the direction of changes in distribution for the European birds during that time period, compiled mainly from publications and personal communications. Unfortunately, Burton (1995) did not cite the exact sources of this information, and the precision of the information is difficult to be evaluated.

Table 1. Predictions on the expansion of the distribution ranges for European bird species with different distributions, according to the hypothesis of a response to climatic warming, and the alternative hypothesis of changes of distribution provoked by other causes.

	Hypothesis 1: Climate warming	Hypothesis 2: Other causes
	Margin advancing	Margin advancing
N	North or nothing	South
CN	North or nothing	South
C	North	Both
SC	North	North
S	North	North
SCN	North or nothing	Both or nothing
Total	North	Both

S: south, SC: south-centre, C: centre, CN: centre-north, N: north, SCN: south-centre-north.

In this chapter, introduced species were excluded (e.g., Mandarin Duck, *Aix galericulata*; Ruddy Duck, *Oxyura jamaicensis*). Some species were taxonomically divided in two after the publication of the Burton's work (e.g., Herring Gull, *Larus argentatus*; Imperial Eagle, *Aquila heliaca*; Grey Shrike, *Lanius excubitor*); for these species, I used the original taxonomy appearing in Burton (1995), and they are included as a species. In the case of subspecies (e.g., Carrion Crow, *Corvus corone corone*, and Hooded Crow, *C. c. cornix*), I considered the distribution change in the whole species. I recorded as having advanced northwards those species that had shifted their distribution margins north, north-east and/or north-west, according to Burton (1995), and southwards those shifting their margins south, south-east and/or south-west. Species that advanced westwards or eastwards, or those expanding northwards as well as southwards, were not considered because data in Burton (1995) are simple and analyses in this sense were unviable. Those species with a doubtful advance or retreat (catalogued as "?" by Burton) were included or omitted according to their particular characteristics. For example, Burton (1995) considered that the Dupont's Lark (*Chersophilus duponti*) could have advanced northwards, but there was doubt with this species (category "N?"). Probably, this species had not expanded northwards, being previously undetected (Garza & Suárez 1990); thus this species was not included as expanding northwards. Species with distribution changes that stopped before 1980 (e.g., Black Stork, *Ciconia nigra,* or White-headed Duck, *Oxyura leucocephala*) were not considered, as the recent climatic warming began in the 70s (Walther *et al.* 2002).

Previously, the map of Europe was divided in three sectors (north, centre and south) by two parallel lines crossing the map at 43° 30' and 57° 30' of latitude north. Each species was categorized according to their distribution in these sectors using the maps in Cramp (1998): south (S), south-centre (SC), centre (C), centre-north (CN), north (N), south-centre-north (SCN) (see Appendix 1 for the category for each species). For each category of distribution, I contrasted the frequency of species that advanced southwards or northwards with a frequency expected (null hypothesis) of equal number of species advancing in each direction, using the Fisher Exact Test.

RESULTS

Out of 425 species considered, 200 species (47.1%) showed a directional advance of their margins northwards (150; 35.3%) or southwards (50; 11.8%) (Table 2). The frequency of species that advanced northwards was significantly higher than expected by the null hypothesis (100 species northwards and 100 southwards; Fisher Exact Test, $P < 0.001$). Considering each category of distribution, 32.8% of southern species and the 47.8% of those distributed in southern-central Europe, showed a northward advance of their northern distribution margins, while the southern margin of no species with these distributions advanced southwards (Table 2). The northward advance was more frequent than expected by chance ($P < 0.001$ for both cases). In contrast, only the 2.5% of northern species showed a northward advance of their northern margins, versus 36 species (44.4%) advancing their southern margins southwards. Therefore, northern species advanced significantly southwards ($P = 0.001$). For the species with central-northern distribution, 17 (27.0%) advanced northwards, versus 13 (20.6%) advancing southwards, which did not differ from the chance ($P = 0.80$). The pattern for the northern species significantly differed from that for the central-northern ones ($P < 0.001$; Table 2).

Table 2. Number (%) of species for each category of distribution and the total that showed northward or southward directional changes in their distribution margins.

	Margin advancing		
	North	Stable	South
N	2 (2.5)	43 (53.1)	36 (44.4)
CN	17 (27.0)	33 (52.4)	13 (20.6)
C	15 (46.9)	17 (53.1)	0 (0.0)
SC	43 (47.8)	47 (52.2)	0 (0.0)
S	20 (32.8)	41 (67.2)	0 (0.0)
SCN	53 (54.1)	44 (44.9)	1 (1.0)
Total	150 (35.3)	225 (52.9)	50 (11.8)

S: south, SC: south-centre, C: centre, CN: centre-north, N: north, SCN: south-centre-north

For the species distributed in central Europe, 46.9% advanced their northern ranges northwards versus 0% advancing their southern ones southwards ($P < 0.01$; Table 2). With respect to the species distributed by south-centre-north, 54.1% advanced their northern distribution range northwards, and only one (1.0%) advanced southwards ($P < 0.001$). There was no difference among distribution categories for the frequency of species changing their distribution ranges ($\chi_5^2 = 7.56$; $P = 0.18$; Table 2).

CONCLUSION

The findings reveal that European bird species have advanced their distributions northward more frequently than southward, this being consistent with a response to climatic warming (Hypothesis 1), although there is also some support for the expansion due to other causes in the northernmost species (Table 3). Southern species (S and SC) have expanded their distribution primarily northward, which was predicted by both hypotheses (Table 1). More significantly, species from central Europe (C) have expanded their distribution exclusively northward (15 northward versus 0 southward), which was predicted only by the hypothesis of the climatic warming, but not by the alternative hypothesis. Moreover, the species distributed throughout the whole of Europe (SCN) have also expanded their distribution clearly northward. In turn, the northernmost species expanded significantly southward (but see Brommer 2004), which was predicted by the alternative hypothesis (expansion due to other causes), but not by the hypothesis of the expansion due to climatic warming. However, two northern species expanded their northern range northward, while no southern, southern-central or even central, and only one southern-central-northern species, showed a southward expansion of their southern ranges. The distribution change of the central-northern species was intermediate between both hypotheses, expanding their ranges equally southward and northward, but it should be noted that the northward advance of these species was probably limited by the Arctic Sea.

Table 3. Support of the results to the predictions in Table 1, with respect to the shifts in distribution ranges in the European birds (Table 2).

	Hypothesis 1: climate warming	Hypothesis 2: other causes
N	Some (2 species northward)	Yes (southward)
CN	Some (no trend southwards)	Perhaps (both directions equal)
C	Yes (northward)	No (0 species southward)
SC	Yes (northward)	Yes (northward)
S	Yes (northward)	Yes (northward)
SCN	Yes (northward)	No (only 1 species southwards)
Total	Yes (northward)	Some

S: south, SC: south-centre, C: centre, CN: centre-north, N: north, SCN: south-centre-north.

These results, as a whole, suggest that some European species varied their distribution by causes other than climatic warming, although, in general, many other species have responded to climatic warming by an expansion northward of their distribution. This pattern for the whole of Europe is consistent with the results found at lower scale for the United Kingdom (Thomas & Lennon 1999, Hickling *et al.* 2006), and Finland (Brommer 2004), and at almost continental scale in United States of America (Hitch & Leberg 2007, La Sorte & Thompson 2007), and with the results found for other organisms (Parmesan *et al.* 1999, Sturm *et al.* 2001, Thomas *et al.* 2001). The distribution of birds is strongly mediated by temperature (Root 1988, Turner *et al.* 1988, Lennon *et al.* 2000), and thus the response capacity of European birds to climatic warming will favour their survival (Huntley *et al.* 2006). However, the northernmost species face a high risk of extinction due to climatic warming, as their capacity to expand northward is limited (Parmesan 2006).

In conclusion, the results in this chapter support for the whole European continent the hypothesis that bird species have advanced their distributions northward in response to climatic warming, and therefore they have a potential to respond to the future climatic warming by changing their distributions.

ACKNOWLEDGMENTS

I am thankful to Jorge Castro, Óscar Gordo, David Nesbitt and Manuel Pizarro for their help in this study.

REFERENCES

Brommer, J. E. (2004). The range margins of northern birds shift polewards. *Annales Zoologici Fennici, 41,* 391-397

Burton, J. F. (1995). *Birds and climate change.* London, United Kingdom: A & C Black.

Carrillo, C. M. Barbosa, A. Valera, F. Barrientos, R. & Moreno, E. (2007). Northward expansion of a desert bird: Effects of climate change? *Ibis, 149,* 166-169.

Cramp, S. (1998). *The complete birds of the Western Palearctic on CD-ROM.* Oxford, United Kingdom: Oxford University Press.

Garza, V. & Suárez, F. (1990). Distribución, población y selección de hábitat de la Alondra de Dupont (*Chersophilus duponti*) en la Península Ibérica. *Ardeola, 37,* 3-12.

Hickling, R. Roy, D. B. Hill, J. K. Fox, R. & Thomas, C. D. (2006). The distributions of a wide range of taxonomic groups are expanding polewards. *Global Change Biology, 12,* 450-455.

Hitch, A. L. & Leberg, P. L. (2007). Breeding distributions of North American bird species moving North as a result of climate change. *Conservation Biology, 21,* 534-539.

Hill, J. K. Collingham, Y. C. Thomas, C. D. Blakeley, D. S. Fox, R. Moss, D. & Huntley, B. (2001). Impacts of landscape structure on butterfly range expansion. *Ecology Letters, 4,* 313-321.

Houghton, J. (2004). *Global warming* (3rd edition). Cambridge, United Kingdom: Cambridge University Press.

Hughes, L. (2000). Biological consequences of global warming: is the signal already. *Trends in Ecology and Evolution, 15,* 56-61.

Huntley, B. Collingham, Y. C. Green, R. E. Hilton, G. M. Rahbek, C. & Willis, S. G. (2006). Potential impacts of climatic change upon geographical distribution of birds. *Ibis, 148,* 8-28.

IPCC. (2001). *Climate change 2001: The scientific basis*. Cambridge, United Kingdom: Cambridge University Press.

La Sorte, F. A. & Thompson III, F. R. (2007). Poleward shifts in winter ranges of North American birds. *Ecology, 88,* 1803-1812.

Lennon, J. J. Greenwood, J. J. D. & Turner, J. R. G. (2000). Bird diversity and environmental gradients in Britain: a test of the species-energy hypothesis. *Journal of Animal Ecology, 69,* 581-598.

Parmesan, C. (1996). Climate and species' range. *Nature, 382,* 765-766.

Parmesan, C. (2006). Ecological and evolutionary responses to recent climate change. *Annual Review of Ecology, Evolution and Systematics, 37,* 637-669.

Parmesan, C. Ryrholm, N. Stefanescu, C. Hill, J. K. Thomas, C. D. Descimon, H.; Huntley, B. Kaila, L. Kullberg, J. Tammaru, T. Tennent, W. J. Thomas, J. A. & Warren, M. (1999). Poleward shifts in geographical ranges of butterfly species associated with regional warming. *Nature, 399,* 579-583.

Parmesan, C. & Yohe, G. (2003). A globally coherent fingerprint of climate change impacts across natural systems. *Nature, 421,* 37-42.

Root, T. (1988). Energy constraints on avian distributions and abundances. *Ecology, 69,* 330-339.

Root, T. L. Price, J. T. Hall, K. R. Schneider, S. H. Rosenzweig, C. & Pounds, J. A. (2003). Fingerprints of global warming on wild animals and plants. *Nature, 421,* 57-60.

Sanz, J. J. (2002). Climate change and birds: have their ecological consequences already been detected in the Mediterranean region? *Ardeola, 49,* 109-120.

Sturm, M. Racine, C. & Tape, K. (2001). Increasing shrub abundance in the Arctic. *Nature, 411,* 546-547.

Thomas, C. D. Bodsworth, E. J. Wilson, R. J. Simmons, A. D. Davies, Z. G. Musche, M. & Conradt, L. (2001). Ecological and evolutionary processes at expanding range margins. *Nature, 411,* 577-581.

Thomas, C. D. Cameron, A. Green, R. E. Bakkenes, M. Beaumont, L. J. Collingham, Y. C. Erasmus, B. F. N. Ferreira de Siqueira, M. Grainger, A. Hannah, L. Hughes, L. Huntley, B. van Jaarsveld, A. S. Midgley, G. F. Miles, L. Ortega-Huerta, M. A. Peterson, A. T. Phillips, O. L. & Williams, S. E. (2004). Extinction risk from climate change. *Nature, 427,* 145-148.

Thomas, C. D. & Lennon, J. J. (1999). Birds extend their ranges northwards. *Nature, 399,* 213-213.

Turner, J. R. G. Lennon, J. J. & Lawrenson, J. A. (1988). British bird species distributions and the energy theory. *Nature, 335,* 539-541.

Walther, G. R. Post, E. Convey, P. Menzel, A. Parmesan, C. Beebee, T. J. C. Fromentin, J. M. Hoegh-Guldberg, O. & Bairlein, F. (2002). Ecological responses to recent climate change. *Nature, 416,* 389-395.

Appendix 1. List of bird species used in this study, showing the distribution margin (northern or southern) that has expanded, as well as the distribution category assigned.

English name	Scientific name	Category	Advance
Red-throated diver	*Gavia stellata*	N	S
Black-throated diver	*Gavia arctica*	N	S
Great northern diver	*Gavia immer*	N	S
White-billed diver	*Gavia adamsii*	N	S
Little grebe	*Tachybaptus ruficollis*	SC	N
Great crested grebe	*Podiceps cristatus*	SC	N
Red-necked grebe	*Podiceps grisegena*	C	
Slavonian grebe	*Podiceps auritus*	N	S
Black-necked grebe	*Podiceps nigricollis*	SC	N
Fulmar	*Fulmarus glacialis*	CN	S
Cory's shearwater	*Calonectris diomedea*	S	
Manx shearwater	*Puffinus puffinus*	SCN	
Storm petrel	*Hydrobates pelagicus*	SCN	
Leach's storm petrel	*Oceanodroma leucorhoa*	N	
Gannet	*Morus bassanus*	N	N
Cormorant	*Phalacrocorax carbo*	SCN	
Shag	*Phalacrocorax aristotelis*	SCN	N
Pygmy cormorant	*Phalacrocorax pygmeus*	S	N
White pelican	*Pelecanus onocrotalus*	C	
Dalmatian pelican	*Pelecanus rufescens*	C	
Bittern	*Botaurus stellaris*	SC	N
Little bittern	*Ixobrichus minutus*	SC	
Night heron	*Nycticorax nycticorax*	SC	N
Squacco heron	*Ardeola ralloides*	SC	N
Cattle egret	*Bubulcus ibis*	S	N
Little egret	*Egretta garzetta*	S	N
Grey white egret	*Egretta alba*	C	N
Grey heron	*Ardea cinerea*	SCN	N
Purple heron	*Ardea purpurea*	SC	
Black stork	*Ciconia nigra*	SC	
White stork	*Ciconia ciconia*	SC	
Glossy ibis	*Plegadis falcinellus*	S	
Spoonbill	*Platalea leucorodia*	SC	
Greater flamingo	*Phoenicopterus ruber*	S	
Mute swan	*Cygnus olor*	C	N
Bewick's swan	*Cygnus bewickii*	N	
Whooper swan	*Cygnus cygnus*	N	S
Bean goose	*Anser fabalis*	N	
Pink-footed goose	*Anser brachyrhynchus*	N	
White-fronted goose	*Anser albifrons*	N	
Lesser white-fronted goose	*Anser erythropus*	N	
Greylag goose	*Anser anser*	CN	S
Canada goose	*Branta canadensis*	CN	N
Barnacle goose	*Branta leucopsis*	N	S
Brent goose	*Branta bernicla*	N	
Ruddy shelduck	*Tadorna ferruginea*	SC	
Shelduck	*Tadorna tadorna*	SCN	N
Wigeon	*Anas penelope*	N	
Gadwall	*Anas strepera*	C	N
Teal	*Anas crecca*	CN	
Mallard	*Anas platyrhynchos*	SCN	N

Pintail	*Anas acuta*	CN	
Garganey	*Anas querquedula*	CN	
Shoveler	*Anas clypeata*	CN	
Marbled teal	*Marmaronetta angustirostris*	S	N
Red-crested pochard	*Netta rufina*	SC	N
Pochard	*Aythya ferina*	C	N
Ferruginous duck	*Aythya niroca*	SCN	
Tufted duck	*Aythya fuligula*	CN	N
Scaup	*Aythya marila*	N	S
Eider	*Somateria mollissima*	N	S
King eider	*Somateria spectabilis*	N	S
Steller's eider	*Polysticta stelleri*	N	S
Harlequin duck	*Histrionicus histrionicus*	N	
Long-tailed duck	*Clangula hyemalis*	N	S
Common scoter	*Melanitta nigra*	N	
Velvet scoter	*Melanitta fusca*	N	
Barrow's goldeneye	*Bucephala islandica*	N	
Goldeneye	*Bucephala clangula*	N	S
Smew	*Mergellus albellus*	N	S
Red-breasted merganser	*Mergus serrator*	N	S
Goosander	*Mergus merganser*	CN	S
White-headed duck	*Oxyura leucocephala*	SC	N
Honey buzzard	*Pernis apivorus*	SCN	N
Black-winged kite	*Elanus caeruleus*	S	N
Black kite	*Milvus migrans*	SCN	N
Red kite	*Milvus milvus*	SC	N
White-tailed eagle	*Haliaeetus albicilla*	SCN	
Lammergeier	*Gypaetus barbatus*	S	
Egyptian vulture	*Neophron percnopterus*	S	
Griffon vulture	*Gyps fulvus*	S	
Black vulture	*Aegypius monachus*	S	
Short-toed eagle	*Circaetus gallicus*	SC	
Marsh harrier	*Circus aeruginosus*	SCN	N
Hen harrier	*Circus cyaneus*	CN	S
Pallid harrier	*Circus macrourus*	C	
Montagu's harrier	*Circus pygargus*	SC	N
Goshawk	*Accipiter gentilis*	SCN	
Sparrowhawk	*Accipiter nisus*	SCN	
Levant sparrowhawk	*Accipiter brevipes*	SC	N
Buzzard	*Buteo buteo*	SCN	
Long-legged buzzard	*Buteo rufinus*	S	
Rough-legged buzzard	*Buteo lagopus*	N	
Lesser spotted eagle	*Aquila pomarina*	C	
Spotted eagle	*Aquila clanga*	CN	
Steppe eagle	*Aquila rapax*	C	
Imperial eagle	*Aquila heliaca*	S	N
Golden eagle	*Aquila chrysaetos*	SCN	
Booted eagle	*Hieraaetus pennatus*	SC	
Bonelli's eagle	*Hieraaetus fasciatus*	S	
Osprey	*Pandion haliaetus*	SCN	S
Lesser kestrel	*Falco naumanni*	SC	N
Kestrel	*Falco tinnuculus*	SCN	
Red-footed falcon	*Falco vespertinus*	C	N
Merlin	*Falco columbarius*	N	
Hobby	*Falco subbuteo*	SCN	N

Appendix 1. (Continued)

Eleanora's falcon	*Falco eleonorae*	S	
Lanner falcon	*Falco biarmicus*	S	
Saker falcon	*Falco cherrug*	C	N
Gyrfalcon	*Falco rusticolus*	N	S
Peregrine	*Falco peregrinus*	SCN	
Hazel hen	*Bonasa bonasia*	CN	
Red grouse	*Lagopus lagopus*	N	
Ptarmigan	*Lagopus mutus*	CN	
Black grouse	*Lyrurus tetrix*	CN	
Capercaillie	*Tetrao urogallus*	CN	
Chukar partridge	*Alectoris chukar*	S	
Rock partridge	*Alectoris graeca*	SC	
Red-legged partridge	*Alectoris rufa*	SC	
Barbary partridge	*Alectoris barbara*	S	
Grey partridge	*Perdiz perdix*	SCN	
Quail	*Coturnix coturnix*	SC	N
Pheasant	*Phasianus colchicus*	SC	
Andalusian hemipode	*Turnix sylvatica*	S	
Water rail	*Rallus aquaticus*	SCN	N
Spotted crake	*Porzana porzana*	SCN	N
Little crake	*Porzana parva*	SC	
Baillon's crake	*Porzana pusilla*	SC	
Corncrake	*Crex crex*	CN	
Moorhen	*Gallinula chloropus*	SC	N
Purple gallinule	*Porphyrio porphyrio*	S	
Coot	*Fulica atra*	SC	N
Crested coot	*Fulica cristata*	S	
Crane	*Grus grus*	CN	
Demoiselle crane	*Anthropoides virgo*	C	
Little bustard	*Tetrax tetrax*	SC	
Great bustard	*Otis tarda*	SC	
Oystercatcher	*Haematopus ostralegus*	SCN	N
Black-winged stilt	*Himantopus himantopus*	SC	N
Avocet	*Recurvirostra avosetta*	SC	N
Stone curlew	*Burhinus oedicnemus*	SC	
Collared pratincole	*Glareola pratincola*	S	
Black-winged pratincole	*Glareola nordmanni*	C	
Little ringed plover	*Charadrius dubius*	SCN	
Ringed plover	*Charadrius hiaticula*	N	
Kentish plover	*Charadrius alexandrinus*	SC	
Dotterel	*Charardrius morinellus*	CN	S
Golden plover	*Pluvialis apricaria*	N	S
Grey plover	*Pluvialis squatarola*	N	
Lapwing	*Vanellus vanellus*	SCN	N
Knot	*Calidris tenuirostris*	N	
Sanderling	*Calidris alba*	N	
Little stint	*Calidris minuta*	N	
Temminck's stint	*Calidris temminckii*	N	S
Purple sandpiper	*Calidris maritima*	N	
Dunlin	*Calidris alpina*	N	
Broad-billed sandpiper	*Limicola falcinellus*	N	S
Ruff	*Philomachus pugnax*	CN	N
Jack snipe	*Lymnocryptes minimus*	N	S

European Bird Species have Expanded Northwards During ...

Snipe	*Gallinago gallinago*	CN	
Great snipe	*Gallinago media*	CN	
Woodcock	*Scolopax rusticola*	CN	N
Black-tailed godwit	*Limosa limosa*	CN	N
Bar-tailed godwit	*Limosa lapponica*	N	
Whimbrel	*Numenius phaeopus*	N	S
Curlew	*Numenius arquata*	CN	N
Spotted redshank	*Tringa erythropus*	N	
Redshank	*Tringa totanus*	SCN	
Marsh sandpiper	*Tringa stagnatilis*	C	N
Greenshank	*Tringa nebularia*	N	S
Green sandpiper	*Tringa achropus*	CN	
Wood sandpiper	*Tringa glareola*	CN	S
Terek sandpiper	*Xenus cinereus*	CN	
Common sandpiper	*Actitis hypoleucos*	SCN	
Turnstone	*Arenaria interpres*	N	S
Red-necked phalarope	*Phalaropus lobatus*	N	
Grey phalarope	*Phalaropus fulicarius*	N	
Pomarine skua	*Stercorarius pomarinus*	N	N
Arctic skua	*Stercorarius parasiticus*	N	S
Long-tailed skua	*Stercorarius longicaudus*	N	S
Great skua	*Catharacta skua*	N	
Great black-headed gull	*Larus ichthyaetus*	S	
Mediterranean gull	*Larus melanocephalus*	S	N
Little gull	*Larus minutus*	CN	N
Sabine's gull	*Larus sabini*	N	
Black-headed gull	*Larus ridibundus*	CN	N
Slender-billed gull	*Larus genei*	S	
Audouin's gull	*Larus audouinii*	S	
Common gull	*Larus canus*	N	
Lesser black-backed gull	*Larus fuscus*	CN	N
Herring gull	*Larus argentatus*	SCN	N
Glaucous gull	*Larus hyperboreus*	N	
Great black-backed gull	*Larus marinus*	CN	N
Kittiwake	*Rissa tridactyla*	CN	S
Ivory gull	*Pagophila eburnea*	N	S
Gull-billed tern	*Gelochelidon nilotica*	SC	N
Caspian tern	*Sterna caspia*	CN	
Sandwich tern	*Sterna sandvicensis*	SC	
Roseate tern	*Sterna dougallii*	CN	
Common tern	*Sterna hirundo*	SCN	N
Arctic tern	*Sterna paradisaea*	CN	
Little tern	*Sterna albifrons*	SC	N
Whiskered tern	*Chlidonias hybridus*	SC	N
Black tern	*Chlidonias niger*	SC	N
White-winged black tern	*Chlidonias leucopterus*	C	
Guillemot	*Uria aalge*	CN	S
Brünnich's guillemot	*Uria lomvia*	N	
Razorbird	*Alca torca*	CN	
Black guillemot	*Cepphus grylle*	N	
Little auk	*Alle alle*	N	S
Puffin	*Fratercula arctica*	CN	S
Black-bellied sandgrouse	*Pterocles orientalis*	S	
Pin-tailed sandgrouse	*Pterocles alchata*	S	
Rock dove	*Columba livia*	SC	

Appendix 1. (Continued)

Stock dove	Columba oenas	SCN	
Wood pigeon	Columba palumbus	SCN	N
Collared dove	Streptopelia decaocto	C	
Turtle dove	Streptopelia turtur	SC	
Palm dove	Streptopelia senegalensis	S	N
Great spotted cuckoo	Clamator glandarius	S	N
Cuckoo	Cuculus canorus	SCN	
Oriental cuckoo	Cuculus saturatus	CN	
Barn owl	Tyto alba	SC	
Scops owl	Otus scops	SC	N
Eagle owl	Bubo bubo	SCN	N
Snowy owl	Nyctea scandiaca	N	S
Hawk owl	Surnia ulula	N	
Pygmy owl	Glaucidium passerinum	CN	
Little owl	Athene noctua	SC	
Tawny owl	Strix aluco	SCN	N
Ural owl	Strix uralensis	CN	
Great grey owl	Strix nebulosa	N	S
Long-eared owl	Asio otus	SCN	N
Short-eared owl	Asio flammeus	CN	N
Tengmalm's owl	Aegolius funereus	CN	N
Nightjar	Caprimulgus europaeus	SCN	N
Red-necked nightjar	Caprimulgus ruficollis	S	
Swift	Apus apus	SCN	
Pallid swift	Apus pallidus	S	N
Alpine swift	Apus melba	SC	N
White-rumped swift	Apus caffer	S	N
Little swift	Apus affinis	S	N
Kingfisher	Alcedo atthis	SC	
Bee-eater	Merops apiaster	SC	N
Roller	Coracias garrulus	SC	
Hoopoe	Upupa epops	SC	
Wryneck	Jynx torquilla	SCN	
Grey-headed woodpecker	Picus canus	CN	
Green woodpecker	Picus viridis	SCN	N
Black woodpecker	Dryocopus martius	SCN	
Great spotted woodpecker	Dendrocopos major	SCN	N
Syrian woodpecker	Dendrocopos syriacus	SC	N
Middle spotted woodpecker	Dendrocopos medius	SC	
White-backed woodpecker	Dendrocopos leucotos	CN	
Lesser spotted woodpecker	Dendrocopos minor	SCN	N
Three-toed woodpecker	Picoides tridactylus	CN	
Dupont's lark	Chersophilus duponti	S	
Calandra lark	Melanocorypha calandra	SC	
Short-toed lark	Calandrella brachydactila	SC	
Lesser short-toed lark	Calandrella rufescens	SC	
Crested lark	Galerida cristata	SC	
Thekla lark	Galerida theklae	S	
Woodlark	Lullula arborea	SCN	N
Skylark	Alauda arvensis	SCN	
Shore lark	Emerophila alpestris	SCN	N
Sand martin	Riparia riparia	SCN	
Crag martin	Tptyonoprogne rupestris	SC	N

European Bird Species have Expanded Northwards During … 149

Swallow	*Hirundo rustica*	SCN	N
Red-rumped swallow	*Hirundo daurica*	S	N
House martin	*Delichon urbica*	SCN	
Tawny pipit	*Anthus campestris*	SC	N
Tree pipit	*Anthus trivialis*	SCN	
Meadow pipit	*Anthus pratensis*	CN	
Red-throated pipit	*Anthus cervinus*	N	S
Rock pipit	*Anthus spinoletta*	SC	
Yellow wagtail	*Motacilla flava*	SCN	N
Citrine wagtail	*Motacilla citreola*	C	
Grey wagtail	*Motacilla cinerea*	SC	N
White wagtail	*Motacilla alba*	SCN	N
Waxwing	*Bombycilla garrulus*	N	
Dipper	*Cinclus cinclus*	SCN	
Wren	*Troglodytes troglodytes*	SCN	
Dunnock	*Prunella modularis*	SCN	N
Alpine acentor	*Prunella collaris*	SC	
Rufous bushchat	*Cercotrichas galactotes*	S	
Robin	*Erithacus rubecula*	SCN	
Thrush nightingale	*Luscinia luscinia*	C	N
Nightingale	*Luscinia megarhynchos*	SC	
Bluethroat	*Luscinia svecica*	CN	S
Red-flanked bluetail	*Tarsiger cyanurus*	N	
Black redstart	*Phoenicurus ochruros*	SC	N
Redstart	*Phoenicurus phoenicurus*	SCN	
Whinchat	*Saxicola rubetra*	CN	
Stonechat	*Saxicola torquata*	SC	
Isabelline wheatear	*Oenanthe isabellina*	S	
Wheatear	*Oenanthe oenanthe*	SCN	
Pied wheatear	*Oenanthe pleschanka*	C	
Black-eared wheatear	*Oenanthe hispanica*	S	N
Black wheatear	*Oenanthe leucura*	S	
Rock thrush	*Monticola saxatilis*	SC	
Blue rock thrush	*Monticola solitarius*	S	
Ring ouzel	*Turdus torquatus*	CN	
Blackbird	*Turdus merula*	SCN	N
Fieldfare	*Turdus pilaris*	CN	
Song thrush	*Turdus philomelos*	SCN	
Redwing	*Turdus iliacus*	N	S
Mistle thrush	*Turdus viscivorus*	SCN	
Cetti's warbler	*Cettia cetti*	SC	N
Fan-tailed warbler	*Cisticola juncidis*	S	N
Lanceolated warbler	*Locustella lanceolata*	N	
Grasshopper warbler	*Locustella naevia*	C	N
River warbler	*Locustella fluviatilis*	C	N
Savi's warbler	*Locustella luscinioides*	SC	N
Moustached warbler	*Acrocephalus melanopogon*	SC	
Aquatic warbler	*Acrocephalus paludicola*	C	
Sedge warbler	*Acrocephalus schoenobaenus*	SCN	N
Paddyfield warbler	*Acrocephalus agricola*	C	N
Blyth's reed warbler	*Acrocephalus dumetorum*	CN	
Marsh warbler	*Acrocephalus palustris*	C	N
Reed warbler	*Acrocephalus scirpaceus*	SC	N
Great reed warbler	*Acrocephalus arundinaceus*	SC	N
Olivaceous warbler	*Hippolais pallida*	S	N

Appendix 1. (Continued)

Booted warbler	*Hippolais caligata*	C	N
Olive-tree warbler	*Hippolais olivetorum*	S	N
Icterine warbler	*Hippolais icterina*	CN	N
Melodious warbler	*Hippolais polyglotta*	SC	N
Marmora's warbler	*Sylvia sarda*	S	
Dartford warbler	*Sylvia undata*	SC	N
Spectacled warbler	*Sylvia conspicillata*	S	
Subalpine warbler	*Sylvia cantillans*	S	
Sardinian warbler	*Sylvia melanocephala*	S	
Rüppell's warbler	*Sylvia rüppelli*	S	
Orphean warbler	*Sylvia hortensis*	SC	
Barred warbler	*Sylvia nisoria*	C	N
Lesser whitethroat	*Sylvia curruca*	SCN	N
Whitethroat	*Sylvia communis*	SCN	
Garden warbler	*Sylvia borin*	SCN	
Blackcap	*Sylvia atricapilla*	SCN	N
Greenish warbler	*Phylloscopus trochiloides*	CN	N
Arctic warbler	*Phylloscopus borealis*	N	
Bonelli's warbler	*Phylloscopus bonelli*	SC	N
Wood warbler	*Phylloscopus sibilatrix*	SCN	N
Chiffchaff	*Phylloscopus collybita*	SCN	N
Willow warbler	*Phylloscopus trochilus*	CN	
Goldcrest	*Regulus regulus*	SCN	N
Firecrest	*Regulus ignicapillus*	SC	N
Spotted flycatcher	*Muscicapa striata*	SCN	N
Red-breasted flycatcher	*Ficedula parva*	C	N
Collared flycatcher	*Ficedula albicollis*	SC	N
Pied flycatcher	*Ficedula hypoleuca*	SCN	N
Bearded tit	*Panurus biarmicus*	SC	
Long-tailed tit	*Aegithalus caudatus*	SCN	
Marsh tit	*Parus palustris*	SCN	N
Sombre tit	*Parus lugubris*	S	
Willow tit	*Parus montanus*	CN	
Siberian tit	*Parus cinctus*	N	S
Crested tit	*Parus cristatus*	SCN	N
Coal tit	*Parus ater*	SCN	N
Blue tit	*Parus caeruleus*	SCN	N
Azure tit	*Parus cyanus*	C	
Great tit	*Parus major*	SCN	N
Corsican nuthatch	*Sitta whiteheadi*	S	
Nuthatch	*Sitta europaea*	SCN	N
Rock nuthatch	*Sitta neumayer*	S	
Wallcreeper	*Tichodroma muraria*	C	
Treecreeper	*Certhia familiaris*	SCN	
Short-toed treecreeper	*Certhia brachydactila*	SC	N
Penduline tit	*Remiz pendulinus*	SC	N
Golden oriole	*Oriolus oriolus*	SC	N
Red-backed shrike	*Lanius collurio*	SCN	
Lesser grey shrike	*Lanius minor*	SC	
Great grey shrike	*Lanius excubitor*	SCN	
Woodchat shrike	*Lanius senator*	SC	
Masked shrike	*Lanius nubicus*	S	
Jay	*Garrulus glandarius*	SCN	N
Siberian jay	*Perisoreus infaustus*	N	

European Bird Species have Expanded Northwards During ...

Azure-winged magpie	*Cyanopica cyana*	S	
Magpie	*Pica pica*	SCN	N
Nutcraker	*Nucifraga caryocatactes*	CN	N
Chough	*Pyrrhocorax pyrrhocorax*	SC	
Alpine chough	*Pyrrhocorax graculus*	SC	
Jackdaw	*Corvus monedula*	SCN	N
Rook	*Corvus frugilegus*	CN	N
Carrion crow	*Corvus corone*	SCN	N
Raven	*Corvus corax*	SCN	
Starling	*Sturnus vulgaris*	SCN	
Spotless starling	*Sturnus unicolor*	S	N
Rose-coloured starling	*Sturnus roseus*	C	
House sparrow	*Passer domesticus*	SCN	
Spanish sparrow	*Passer hispaniolensis*	S	N
Tree sparrow	*Passer montanus*	SCN	N
Rock sparrow	*Petronia petronia*	S	
Snowfinch	*Montifringilla nivalis*	C	
Chaffinch	*Fringilla coelebs*	SCN	N
Brambling	*Fringilla montifringilla*	N	S
Serin	*Serinus serinus*	SC	N
Citril finch	*Serinus citrinella*	SC	
Greenfinch	*Carduelis chloris*	SCN	N
Goldfinch	*Carduelis carduelis*	SCN	N
Siskin	*Carduelis spinus*	CN	S
Linnet	*Carduelis cannabina*	SCN	
Twite	*Carduelis flavirostris*	CN	S
Redpoll	*Carduelis flammea*	CN	S
Arctic redpoll	*Acanthis hornemanni*	N	
White-winged crossbill	*Loxia leucoptera*	N	
Crossbill	*Loxia curvirostra*	SCN	
Parrot crossbill	*Loxia pytyopsittacus*	N	
Trumpeter finch	*Bucanetes githagineus*	S	N
Scarlet rosefinch	Carpodacus erythrinus	CN	N
Pine grosbeak	*Pinicola enucleator*	N	
Bullfinch	*Pyrrhula pyrrhula*	SCN	
Hawfinch	*Coccothraustes coccothraustes*	SC	
Lapland bunting	*Calcarius lapponicus*	N	S
Snow bunting	*Plectrophenax nivalis*	N	S
Yellowhammer	*Emberiza citrinella*	SCN	
Cirl bunting	*Emberiza cirlus*	SC	
Rock bunting	*Emberiza cia*	SC	
Ortolan bunting	*Emberiza hortulana*	SCN	
Cretzschmar's bunting	*Emberiza caesia*	S	
Rustic bunting	*Emberiza rustica*	N	S
Little bunting	*Emberiza pusilla*	N	
Yellow-breasted bunting	*Emberiza aureola*	CN	
Reed bunting	*Emberiza schoeniclus*	SCN	N
Black-headed bunting	*Emberiza melanocephala*	SC	N
Corn bunting	*Miliaria calandra*	SC	

S: south, SC: south-centre, C: centre, CN: centre-north, N: north,
SCN: south-centre-north.

In: Trends in Ornithology Research
Editors: P. K. Ulrich and J. H. Willett, pp. 153-160

ISBN: 978-1-60876-454-9
© 2010 Nova Science Publishers, Inc.

Chapter 6

THE STUDY OF INTERACTIONS BETWEEN BIRDS AND FLOWERS IN THE NEOTROPICS: A MATTER OF POINT OF VIEW

Márcia A. Rocca[1] and Marlies Sazima[2]*

[1]Laboratório de Ecologia, Departamento de Ciências Biológicas (DCB), Universidade Estadual de Santa Cruz (UESC), Campus Soane Nazaré de Andrade, Pavilhão Jorge Amado, Km 16, Rod. Ilhéus-Itabuna, CEP 45.650-000, Ilhéus, Bahia, Brasil
[2]Departamento de Biologia Vegetal, Instituto de Biologia, Caixa Postal 6109, Universidade Estadual de Campinas (Unicamp), CEP 13083-970, Campinas, São Paulo, Brasil

ABSTRACT

Birds are among the main components for plant reproduction in tropical ecosystems, with hummingbirds being the most important vertebrate pollinators in the Neotropics. Flower-visiting birds of another groups (the perching birds) are often considered as parasites of the flower-hummingbird relationships. These birds do not present a high degree of specialization for nectarivory, although nectar should be a very important component of the diet of some groups. Birds usually also visit non-ornithophilous flowers, as they look for resources in flowers adapted to pollination by other animals as well. However, very few studies have focused on non-ornithophilous flowers, which means looking at the whole group of species visited by birds, from the bird's point of view— the *resource approach*. While visiting non-ornithophilous flowers, birds (usually hummingbirds) may act merely like robbers, thieves or even co-pollinators. Therefore, when the aim of the study is pollination, one should not only look for ornithophilous flowers, but also for other possible bird pollinated species, from the flower's point of view—the *pollination approach*. Studies focusing on this last approach are even scarcer at the community level. It is important to realize that the set of *ornithophilous species* are inside the wider set of *pollinated species* by birds, and this one is contained inside the set of *visited species* by birds. Studies that only pick up ornithophilous species from a

---

* Corresponding author: E-mail: roccamarcia@yahoo.com.br

community are not focussing on pollinated species by birds, but rather on a subset of that. Another problem of point of view is that most studies in the Neotropics are ground based, which may reduce sampling of canopy species information. Observation positions within the canopy greatly enhance this kind of study and should be used more often. As flowers pollinated by perching birds may be more common in the canopies of Neotropical forests, perching bird flowers and their visitors and pollinators are underestimated in communities sampled only from the ground, which means that the majority of the studies on bird-flower interactions in Neotropical forests present good lists of bird and plant species, but very incomplete interaction networks. After putting together these different approaches to study the bird-flower interaction network, we could maybe build—with the help of other animal-flower networks—a picture of a combined model of nested compartments to the whole community, connecting all animal-flower networks by interactions of co-pollination or just visits, reinforcing the idea of communities displaying high connectance.

Birds and their Flowers in the Neotropical Region

Birds are an important component for plant reproduction in tropical ecosystems (Snow 1981, Whelan *et al.* 2008), with hummingbirds being the most important vertebrate pollinators in the Neotropics (Bawa 1990). Other flower-visiting species (perching birds from many families in Passeriformes and also in other Orders; see Rocca & Sazima *in press*) were often considered as parasites of the flower-hummingbird relation (Stiles 1981). These birds do not present a high degree of specialization for nectarivory, although nectar may be a very important component of the diet of some groups (Sick 1997). However, the possible role of perching birds as pollinators seems to be neglected in the Neotropics (Rocca & Sazima 2008) and may even be more frequent in forest canopies (Toledo 1977). Although these birds do not rely only on floral resources (Sick 1997), there are plant species in the Neotropical region that do depend solely on them for pollination (see Rocca & Sazima *in press*).

The Resource Approach

Ornithophilous flowers (*sensu* Faegri & van der Pijl 1979) may represent up to 22% of Angiosperms in tropical communities (Stiles 1981, Feinsinger 1983, Bawa 1990, Morellato & Sazima 1992). However, most of the time, birds also visit non-ornithophilous flowers, as they often look for resources in flowers adapted to pollination by other animals. Indeed, very few studies focused also on non-ornithophilous flowers, which means to look at the whole group of species visited by birds (Feinsinger 1976, Stiles 1978, Araujo 1996, Araujo & Sazima 2003, Dziedzioch *et al.* 2003, Rocca-de-Andrade 2006), from the bird's point of view or the *resource approach*.

It is important to remember that floral features are selected not only by good pollinators, but also by excluding undesirable ones (Endress 1994). Birds visit flowers that share important similar characteristics with ornithophilous flowers: a good volume of nectar as resource in tubular-like structures, which can be the corolla, calyx, fused or joined stamens or extra-floral nectaries. This means that birds may visit melittophilous, psychophilous, sphingophilous and chiropterophilous flowers (Figure 1), many of them displaying flowers not quite showy at all (Rocca-de-Andrade 2006). While visiting non-ornithophilous flowers, birds (in most of the cases, hummingbirds) may act merely like robbers (making holes through the corolla tube) or thieves (when using the corolla opening as a legitimate visit, but not touching reproductive organs, e.g., Rocca & Sazima 2006), but they may also pollinate (see below). Therefore, birds visiting flowers may be more common than strictly during pollination interactions.

The Pollination Approach

When birds pollinate flowers that display features of other pollination syndrome, birds may act as co-pollinators (e.g., of psycophilous species in Varassin & Sazima 2000; of sphingophilous species in Wolff *et al.* 2003; of melittophilous species in Freitas *et al.* 2006; of many species and syndromes in Rocca-de-Andrade 2006). Therefore, when the aim of the study is pollination, one should not only look for ornithophilous flowers, but also for other possible bird-pollinated species, from the flower's point of view or the *pollination approach*.

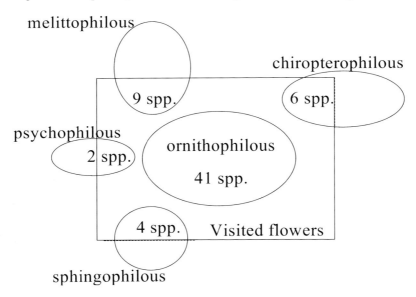

Figure 1. Number of bird-visited species of flowers in an Atlantic rainforest site in South-eastern Brazil (data from Rocca-de-Andrade 2006). Note that in addition to ornithophilous species, birds also visit flowers displaying other pollination syndromes.

It is easier to go to the field and look only for ornithophilous species and study this set of species. But when the question is pollination, which means which plant species share pollinators in place (bird body) or time (plant phenology), one should open up the pool of possible pollinated species. The first step is to take a look at species that share features common to bird-flower ones (Faegri & van der Pijl 1979, Rocca & Sazima *in press*). After this, protocols for determining whether a flower-visitor is a pollinator (Pellmyr 2002) suggest observing if the visitor picks up pollen and if it deposits pollen on co-specific stigma. However, studies focusing on this pollination approach at the community level are even scarcer than those under resource approaches —but are clearly a subset of them.

SET THEORY

It is important to realize that the group of *ornithophilous species* are inside a wider group formed by the *bird-pollinated species*, and this one is a set inside the set of *visited species* by birds, like sets are contained inside other ones (Figure 2). Studies that pick only ornithophilous species from a community are not focussing on pollinated species by birds, but rather on a subgroup of that under the more intense selective pressure of those birds. But from the plant point of view, ornithophilous species and any other species co-pollinated by birds share or compete for these birds like resources— pollen vectors—which means that these species do have something in common.

Ground-based *versus* Canopy-based Studies

Another limitation in studies on bird-flower interactions in the Neotropics is that the great majority of them are ground based, which may reduce sampling of canopy species and mainly of interaction information. As flowers pollinated by perching birds may be more common in the canopy of Neotropical forests than in their understorey (Toledo 1977, Rocca & Sazima 2008), perching bird flowers and their visitors and pollinators are underestimated in communities sampled only by the ground, which means that the majority of the studies on bird-flower interactions in Neotropical forests present good lists of birds and plant species (as birds may be seen or heard from the ground and flowers may fall to the ground), but incomplete interaction networks. Observation positions within the canopy greatly enhance this kind of study of interactions (Nadkarni & Matelson 1989) and should be used more often. Very few studies on birds in the Neotropics made use of it so far (Loiselle 1988, Dulmen 2001, Rocca-de-Andrade 2006, Rocca *et al*. 2006, Canela 2006, Rocca & Sazima 2008). For example, only 9% of bird pollinators of a tree climber species could be seen from the ground in an Atlantic rainforest site, and the presence of the rest of the species would have been completely missed unless observations were made in the canopy (Rocca & Sazima 2008).

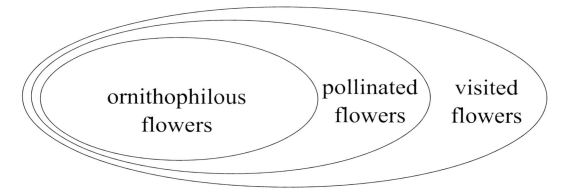

Figure 2. Flowers visited by birds and the subsets of pollinated flowers and ornithophilous flowers.

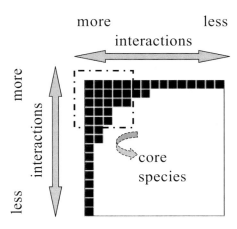

Figure 3. A model of nested animal-plant interaction matrix, showing its main structural features. Black squares standing for recorded interactions between an animal species (in column) and a plant species (in row). Modified from Lewinsohn *et al.* (2006).

Observing All Possible Animal-Plant Interaction Networks

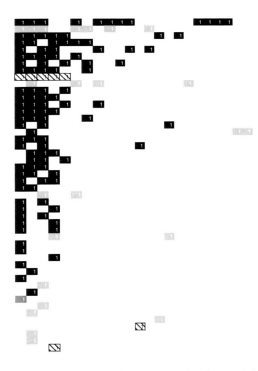

Figure 4. Bird species (in columns) and all visited flower species (in rows) interaction network from an Atlantic rainforest site, SE Brazil (Rocca-de-Andrade 2006) showing different types of interactions: birds and species mainly pollinated by birds (ornithophilous species) in black squares; birds and species mainly pollinated by other animals, being birds secondary pollinators (co-pollination) in gray squares; and birds and species which birds do not pollinate, thus acting like thieves or robbers (simple visit) in dashed squares. Original figure from Rocca-de-Andrade (2006).

If we organize species into pollination matrices, we could see that most of pollination interaction networks are nested (Bascompte *et al.* 2003). This means that, when both species of plants and animals (rows and columns, respectively) are ranked into decreasing number of interactions or from the least specialized ones to the most specialized ones, specialized animals will be only associated to plants with many interactions and plants species with few interactions will be only associated with generalist animal species (Figure 3). Therefore, a dense core of interactions is formed because generalists in one species set tend to interact with generalists in the other (Lewinsohn *et al.* 2006).

As birds visit and even pollinate other species in addition to ornithophilous ones (Figure 4), let us imagine that the network of bird-flower interaction[1] is connected with other animal-flower networks by interactions of co-pollination or just visit, like birds acting as robbers or thieves. These connections generally involve generalist flowers displaying easy access and generalist bird species, linking the core species from both systems. Therefore, the picture of a community network could be a combined model of nested compartments (*sensu* Lewinsohn *et al.* 2006), each one representing a pollination syndrome or system (Figure 5), much more connected than we could previously imagine and contributing to network (and also to the community) robustness or resistance to species loss (*sensu* Bascompte & Jordano 2007).

A broader view of flower visitors and not only of "right pollinators" may elucidate and help to complete the picture that could represent a whole community.

Figure 5. An imaginary combined model of nested compartments of animal-plant interaction networks representing three different pollination syndromes or systems, connected by interactions of pollination or co-pollination (black squares) or just visit (visitors acting as a thief or a robber; white squares) involving core species.

---

[1] Even the so-called bird-flower interaction network may be split into two different ones, as hummingbird and perching bird syndromes display quite different features (see review in Rocca & Sazima *in press*), maybe deserving different networks, also connected by interactions between them.

REFERENCES

Araujo, A. C. (1996). Beija-flores e seus recursos florais numa área de planície costeira do litoral norte de São Paulo. Dissertação de Mestrado, Universidade Estadual de Campinas. 69.

Araujo, A. C. & Sazima, M. (2003). The assemblage of flowers visited by hummingbirds in the "capões" of Southern Pantanal, Mato Grosso do Sul, Brazil. *Flora, 198*, 427-435.

Bascompte, J. & Jordano, P. (2007). Plant-animal mutualistic networks: the architecture of biodiversity. *Annual Review of Ecology, Evolution and Systematics, 38*, 567–93

Bascompte, J. Jordano, P. Melian, C. J. & Olesen, J. M. (2003). The nested assembly of plant-animal mutualistic networks. *Proceedings of the National Academy of Sciences of the United States of America,* 100 *(16),* 9383-9387.

Bawa, K. S. (1990). Plant-pollinator interactions in tropical rain forests. *Annual Review of Ecology and Systematics, 21*, 399-422.

Canela, M. B. F. (2006). Interações entre plantas e beija-flores numa comunidade de floresta Atlântica montana em Itatiaia, R. J. Ph.D Thesis. Universidade Estadual de Campinas (Campinas).

Dulmen, A. V. (2001). Pollination and phenology of flowers in the canopy of two contrasting rain forest types in Amazonia, Colombia. *Plant Ecology, 153*, 73-85.

Dziedzioch, C. Stevens, A. D. & Gottsberger, G. (2003). The hummingbird plant community of a Tropical montane rain forest in Southern Ecuador. *Plant Biology, 5*, 331-337.

Endress, P. K. (1994). *Diversity and evolutionary biology of tropical flowers.* Cambridge University Press, Cambridge. 511 pages.

Faegri, K. & van der Pijl, L. (1979). *Principles of pollination ecology.* 2ª. ed. Pergamon Press. NewYork. 242 pages.

Feinsinger, P. (1976). Organization of a tropical guild of nectarivorous birds. *Ecological Monographs, 46*, 257-291.

Feinsinger, P. (1983). *Coevolution and pollination In*: D. Futuyma & M. Slatkin (Eds.) Coevolution, 282-310. Sinauer Associates, Sunderland, Massachusetts.

Freitas, L. Galetto, L. & Sazima, M. (2006). Pollination by hummingbirds and bees in eight syntopic species and a putative hybrid of Ericaceae in Southeastern Brazil. *Plant Systematics and Evolution, 258*, 49–61.

Lewinsohn, T. M. Prado, P. I. Jordano, P. Bascompte, J. Olesen, J. M. (2006). Structure in plant-animal interaction assemblages. *Oikos, 113(1),* 174-184.

Loiselle, B. A. (1988). Bird abundance and seasonality in a Costa Rican lowland forest canopy. *The Condor, 90*, 761-772.

Morellato, L. P. C. & Sazima, M. (1992). Modos de polinização em uma floresta semidecídua no sudeste do Brasil (Reserva de Santa Genebra, Campinas, SP) *In*: Resumos Seminário Mata de Santa Genebra – conservação e pesquisa em uma reserva florestal urbana em Campinas, Campinas, Brasil. 13.

Nadkarni, N. M. & Matelson, T. J. (1989). Bird use of epiphyte resources in neotropical trees. *The Condor, 91*, 891-907.

Pellmyr, O. (2002). Pollination by animals. p. *In*: Herrera, C. H. & Pellmyr, O. (Eds.) *Plant-animal interactions – An evolutionary approach*, Malden, Blackwell Publishing.

Rocca, M. A. & Sazima, M. (2006). The dioecious, sphingophilous species *Citharexylum myrianthum* (Verbenaceae): Pollination and visitor diversity. *Flora, 201(6),* 440-450.

Rocca, M. A. Sazima, M. & Sazima, I. 2006. Woody woodpecker enjoys soft drinks: the blond-crested woodpecker seeks nectar and pollinates canopy plants in south-eastern Brazil. *Biota Neotropica* 6(2): http://www.biotaneotropica.org.br/v6n2/en/abstract?short-communication+bn02606022006. ISSN 1676-0603

Rocca, M. A. Sazima, M. (2008). Ornithophilous canopy species in the Atlantic rain forest of southeastern Brazil. *Journal Field Ornithology*, *79(2)*, 130-137.

Rocca, M. A. Sazima, M. Beyond hummingbird-flowers: the other side of ornithophily in the Neotropics. *Oecologia Brasiliensis.* In press.

Rocca-de-Andrade, M. A. (2006). Recurso floral para aves em uma comunidade de Mata Atlântica de encosta: sazonalidade e distribuição vertical. Ph.D. Thesis. Universidade Estadual de Campinas (Campinas).

Sick, H. (1997). *Ornitologia Brasileira.* Editora Nova Fronteira, Rio de Janeiro. 912 pages.

Snow, D. W. (1981). *Coevolution of birds and plants. In*: P. L. (Eds.) The evolving biosphere. Part II. Coexistence and coevolution, 169-178. Cambridge University Press, Cambridge, England.

Stiles, F. G. (1978). Temporal organization of flowering among the hummingbird foodplants of a tropical wet forets. *Biotropica 10(3),* 194-210.

Stiles, F. G. (1981). Geographical aspects of bird-flower coevolution, with particular reference to Central America. *Annals of Missouri Botanical Garden, 68*, 323-351.

Toledo, V. M. (1977). Pollination of some rain forest plants by non-hovering birds in Veracruz, Mexico. *Biotropica, 9*, 262-267.

Varassin, I. G. & Sazima, M. (2000). Recursos de Bromeliaceae utilizados por beija-flores e borboletas em Mata Atlântica no Sudeste do Brasil. *Boletim do Museu de Biologia Mello Leitão* 11/12, 57-70.

Whelan, C. J. Wenny, D. G. & Marquis, R. J. 2008 Ecosystem services provided by birds. *Annals of the New York Academy of Sciences* 1134: 25–60.

Wolff, D. Braun, M. & Liede, S. (2003). Nocturnal versus diurnal pollination success in *Isertia laevis* (Rubiaceae): a sphingophilous plant visited by hummingbirds. *Plant Biology, 5*, 71-78.

In: Trends in Ornithology Research
Editors: P. K. Ulrich and J. H. Willett, pp. 161-167

ISBN: 978-1-60876-454-9
© 2010 Nova Science Publishers, Inc.

Chapter 7

RADAR ORNITHOLOGY - THE PAST, PRESENT, AND FUTURE: A PERSONAL VIEWPOINT

Sidney A. Gauthreaux
Banquet Address; Clemson University, Clemson, SC, USA

THE BEGINNING OF RADAR ORNITHOLOGY

Eric Eastwood (March 12, 1910–October 6, 1981) was one of the first to use radar to study the movement of birds, and many of his studies and those of other pioneers are summarized in his book, *Radar Ornithology* published in 1967. He was elected as Fellow of the Royal Society 1968. In the *Biographical Memoirs of Fellows of the Royal Society,* Vol. 29 (November 1983), p. 177–195, F.E. Jones said the following about Eric Eastwood: "An observation that was to prove of great interest to Eastwood in later years was made by operators at a very early CHL type radar station installed at Happisburgh, on the Norfolk coast. Some echoes were positively identified as coming from a flock of geese crossing the sea. This observation, made in 1940, was the first record of the flight of birds being followed by radar and it led to extensive investigations by Eastwood in later years and to the publication of a book on the subject (Eastwood, 1967)." The CHL (Chain Home Low) radar system was developed to counter the low-level air defense threat to the United Kingdom in 1939.

THE WSR-57 YEARS (1957–1993)

In 1957 the first WSR-57 (Weather Surveillance Radar 1957) radars were placed around the northern coast of the Gulf of Mexico as part of a national network of about 50 units. Nine WSR-57 radars were positioned around the Gulf of Mexico from Brownsville, Tex., to Key West, Fla., to monitor the landfall of hurricanes. The systems were very sensitive and could easily detect very light rain in the atmosphere. As a junior in high school with a deep interest in bird migration, I wondered if these powerful weather radars could detect the moisture in the bodies of migrating birds, and soon discovered that on nights when birds were migrating (verified by moon-watching [Lowery, 1951] and flight call counting), the display of the

WSR-57 showed an extensive cloud of snowy targets that disappeared when no migration was underway. I continued to work with the WSR-57 at the New Orleans Weather Bureau during my undergraduate college years (1959-1963) at Louisiana State University in New Orleans (now the University of New Orleans).

During the early 1960s, Frank Bellrose (August 20, 1916–February 19, 2005) of the Illinois Natural History Survey started his investigations of bird migration with the WSR-57 at about the same time that I started my Master of Science thesis research at Louisiana State University in Baton Rouge under the guidance of George H. Lowery, Jr., and Robert J. Newman. I was interested in characterizing the arrival of trans-gulf migrations in the spring and used the WSR-57 at the Lake Charles Weather Bureau in Louisiana for my studies in 1964. While gathering radar data I attempted to identify the sources of the echoes displayed on the radar screen by making observations with a vertically pointing telescope during daylight hours and moon watching at night when the moon was not obscured by cloud during the full-moon period. When the moon was not visible I made observations of migrants passing through the fixed-beam ceilometer at the Lake Charles Weather Bureau. Birds appeared to fly through the beam without deviation on clear nights, and these observations stimulated me to develop a portable ceilometer device that could be used elsewhere on moonless nights (Gauthreaux, 1969). After receiving my M.S. degree in 1965, I continued my studies of the arrival of trans-gulf migration in spring and used the WSR-57 at Lake Charles and New Orleans for my Ph.D. dissertation research at LSU. In addition to gathering data on the diel and seasonal temporal patterns of trans-gulf migration, I also measured the altitudinal distribution and flocking behavior of migrants during daylight and darkness, and quantified the relationship between the amount of reflectivity from migrating birds detected by the radar and the numbers of birds recorded during moon watching (migration traffic rate).

In addition to the quantification of migration using weather surveillance radar, two other discoveries stand out from my dissertation research. The first relates to the geographical locations where trans-gulf migrants stopover after reaching the coast. Prior to my work, Lowery (1945) had postulated that when flying conditions are favorable, trans-gulf migrants continue well inland before landing, creating a geographical zone between the coast and the landing latitude that is empty of trans- gulf migrants—a zone he called the coastal hiatus. Although it was impossible to determine the exact locations where trans-gulf migrants were putting down after flying over the coast and the coastal marshes and prairie because the input of migrants into the stopover areas was a gradual process over several hours, it was obvious from the WSR-57 radar display that the density of echoes thinned dramatically once the latitude of extensive forested land was reached and only rarely did some echoes (flocks of shorebirds?) continue well inland toward central Louisiana before landing. On dates when a trans-gulf flight had arrived in southern Louisiana and weather conditions were favorable for an exodus of migrants, the WSR-57 showed a striking pattern of sudden echo abundance as migrants departed from stopover areas to begin a nocturnal migration 30 to 45 minutes after sunset. For a brief period of time the pattern of echoes from departing migrants delimited the geographical pattern of migration stopover areas within 50 nautical miles of the Lake Charles and New Orleans WSR-57 radars.

The other noteworthy discovery from my dissertation research concerned the 3-dimensional spatial distribution of arriving trans-gulf migrants during daylight and darkness. Trans-gulf migrants generally arrived on the northern gulf coast in species-specific flocks at high altitudes (3,000–6,000 ft above ground level) during the daylight hours, and when arrivals continued into darkness the radar showed that most of the flocks disbanded and the altitude of flight lowered greatly (1,500 ft agl) after dark.

After receiving my doctorate in 1968, I accepted a 2-year post-doctoral fellowship with Eugene P. Odum at the Institute of Ecology at the University of Georgia, continued to use the WSR-57 at Athens, Ga., to study bird migration, and published three papers from my dissertation research (Gauthreaux, 1970, 1971, 1972).

THE ASR-4, 5, AND 7 YEARS (1971–1998)

After joining the faculty at Clemson University in 1970, I began to work with the ASR-4 (Airport Surveillance Radar) at the downtown airport in Greenville, South Carolina. This radar had a moving target indicator so that only moving targets were displayed, and it could detect concentrations of migrating birds out to a range of 60 nautical miles. I could not relate levels of reflectivity to density of birds aloft with the ASR-4 because it did not have variable attenuation (only two levels of sensitivity time control). Instead I compared migration traffic rates (moon-watching and ceilometer techniques) with different patterns of echo density displayed on the radar screen (Gauthreaux, 1974). I continued my work with this unit until the late 1970s when the radar was upgraded (ASR-5) and relocated to the Greenville-Spartanburg Airport in Greer, South Carolina. Several years later (from 1997 through 1998) I used the ASR-7 at Howard Air Force Base in Panama to study the raptor migration during the day and songbird migration at night over the southeastern entrance of the Panama Canal and the northern portion of the Bay of Panama.

MOBILE RADAR LABORATORY DAYS (1980–1 991)

In the late 1970s there was great concern about transmission lines and bird collisions, and I was asked to explore the development of a mobile laboratory with radar that could be used to gather data on the flight characteristics of birds near transmission lines during the day and especially at night. With funding from Electric Power Research Institute (EPRI), I built a mobile radar laboratory that had two configurations of 3-cm (X-band) marine radars (Gauthreaux 1985a, b). One was a commercial, off-the- Shelf, 10-kW marine surveillance radar, and the other was a marine radar with a parabolic dish antenna instead of the typical open array antenna (t-bar) of a marine radar. The stationary antenna of the latter system could be directed at any angle from 90° (horizontal) to 0° (vertical). The horizontal beam configuration was able to monitor a corridor similar to that of the transmission line, and the horizontal surveillance radar could monitor movements approaching and leaving the corridor. A night-vision scope attached to a video camera and binoculars or a telescope were used to visually identify the sources of echoes in the fixed-beam radar at night and during the day, respectively. When told by the radar operator that birds were approaching the line and at what range, field observers using only image intensifiers at night saw three times the number of the birds flying toward a transmission line as did the same observers without knowledge of what the radar was detecting.

The Decline of Neotropical Migrants

In the late 1980s the North American Breeding Bird Survey indicated that populations of many species of Neotropical migrants were in a state of decline. This stimulated me to examine if spring trans-gulf migration might also have changed over a period of years. Because the WSR-57 at Lake Charles had a filmed record archived at the National Climatic Data Center in Asheville, N.C., I was able to compare the radar films from the period when I did my dissertation research (1965–67) to a 3-year period approximately 20 years later (1987–89). Because of the limitation of using the filmed record, I could only record if a trans-gulf flight occurred on a date and could not indicate the peak density of the flight. Nonetheless I was able to determine that the frequency of trans-gulf migrations arriving on the southwestern Louisiana coast had declined by approximately 40 percent over the 20-year period (Gauthreaux, 1992). Could there have been fewer flights of greater magnitude? If so, then there would have been no change in the numbers of trans-gulf migrants. To answer this question, additional WSR57 data were gathered at Slidell, Louisiana, in the early 1990s. I examined the density of flocks on 25- nautical mile range while the radar display on 125-nautical mile range was being archived. I then related the density of flocks to the maximum range of the echo pattern from trans-gulf migration in the archived films and discovered a significant positive relationship between the two (Gauthreaux, 1994). With additional analysis I was able to reject the hypothesis of fewer flights of greater magnitude.

WSR-88D Years (1992–Present)

During the 1980s a major event was underway that would have a great impact on the field of radar ornithology in the United States. In the mid-1980s, Ron Larkin began working with Doppler weather surveillance radar (CHILL) in Illinois that was the prototype for the next generation of weather radar (NEXRAD). The new weather surveillance radar called the WSR-88D (weather surveillance radar, 1988, Doppler) was first deployed and commissioned in the early 1990s, with one at Oklahoma City, Okla. (1990), and one at Melbourne, Fla. (1991). Carroll G. Belser and I began our work with the WSR-88D in the spring of 1992 when the first unit was placed on the northern coast of the Gulf of Mexico at Dickinson, Tex., south of Houston. The WSR-88D is more powerful and much more sensitive than the WSR-57. It also has a 1° beam compared with the 2° beam of the WSR-57, and most importantly, it is a Doppler radar. The radar has three fundamental moments: reflectivity, radial velocity, and spectrum width.

For the next few years, we (Belser and I) validated the WSR-88D with respect to bird movements and quantified the base reflectivity displays of bird migration on the WSR-88D radar by comparing bird density data gathered by moon watching with data on maximum relative reflectivity (dBZ) and reflectivity values (Z) in base reflectivity scans at 0.5° antenna tilt (Gauthreaux and Belser, 1998, 1999a, 2003a). In addition to our work on the arrival of trans-gulf migration (Gauthreaux and Belser, 1999b; Gauthreaux and others, 2006), we have used the WSR-88D to quantify and map the post- breeding roosts of purple martins (*Pro gne subis*) in the Eastern United States (Russell and Gauthreaux, 1998, 1999; Russell and others, 1998). Although much of our work involves migration studies using data from individual

WSR-88 stations, we also monitor nocturnal bird migration nationwide with the national network of 154 WSR-66D radars (Gauthreaux and others, 2003). Because the reflectivity resolution cells of the WSR-88D are 1° x 1 km and smaller than those of the WSR-57 (2° x 1.2 km), the process of using the WSR-88D to delimit migration stopover areas is greatly facilitated. Several studies are underway to identify important migration stopover areas and characterize the habitat so that these resting and refueling areas can be protected (Gauthreaux and Belser, 2003b).

HIGH-RESOLUTION MARINE RADAR (1998–PRESENT)

Although I used a 12-kW marine radar in a mobile laboratory for studies of bird movements near transmission lines in the 1980s, in 1998 I developed a mobile unit based on a 50-kW marine radar with a 1-m-diameter parabolic antenna (2.5° conical beam). This unit was flown by the U.S. Air Force to Howard Air Force Base in Panama where I assessed its capabilities for detecting birds and for helping base operations in their bird/aircraft strike hazard program. The unit readily detected flocks of migrating raptors (broad-winged hawks [*Buteo platypterus*]; Swainson's hawks [*Buteo swainsoni*]; and turkey vultures [*Cathartes aura*]) out to a range of 14 km and individual migrating songbirds at night out to a range of 4–5 km. On the northern coast of the Gulf of Mexico, the unit enabled the detection of flocks of arriving trans-gulf migrants in spring at altitudes up to 12,000–15,000 ft (3,657.6–4,572 m) above ground level. It has also been used to monitor raptor migration through the Rio Grande Basin in south Texas in spring. The echo-trail feature of this radar clearly indicates moving targets, and it is possible to measure the ground speeds of individual targets using the target tracking feature. When the antenna is tilted 30° above the horizontal, the altitude of a target in the radar beam is 1/2 its range; so with this unit it is possible to measure the altitude, flight direction, and ground speed of a target.

THERMAL IMAGING AND VERTICALLY POINTING FIXED-BEAM RADAR (1996–PRESENT)

When using an image intensifier, there is always the possibility that the vertically pointing light beam used to illuminate the underside of the migrants aloft might influence their flight direction and attract them to the light beam. To circumvent this possibility we explored the use of a thermal imaging camera. The camera detects the thermal signature of a bird as it flies over, so no source of illumination is needed. For observing migrants aloft we used the thermal camera with a telephoto lens (4.8° field of view in vertical dimension). To determine the altitude of the birds observed flying through the field of view, we used a 5-kW 3-cm wavelength marine radar with a fixed, vertically pointing parabolic antenna (~ 4.0° conical beam). I used a video screen splitter to combine the video image of the display of the fixed-beam radar and video from the thermal imaging camera and used a date and time generator to label the resultant display. The video was recorded on mini-DV digital video tapes. I analyzed video tapes with the aid of a device that makes time exposures (tracks) of the moving targets in the video record. By looking at the tracks it is possible to distinguish birds from insects and obtain accurate numbers and tracks of individuals within the sample volume (circle = 4.82° observation cone). The video frames with tracks can be saved and enhanced to maximize detection of weak radiance signals from high-flying birds. More details are found in Gauthreaux and Livingston (2006).

THE FUTURE

The future of radar ornithology will benefit greatly from technological developments related to digital processing and communications. Digital processing of raw radar data reflected from targets and captured with marine radar shows great promise. It is possible to obtain automatically quantitative information on the reflectivity of a target and its variation, target flight direction and velocity relative to wind direction and speed, and target track (for example, Nohara and others, 2005). The technology can also provide information on wing-beat patterns of a target passing through a fixed, narrow, conical beam, and this can be used to help identify the source of the radar echo (for example, insect, shorebird, passerine). Currently it is impossible to distinguish a migrating bat from a migrating bird of similar size, so we must be careful when reporting the identification of radar targets without visual or acoustical confirmation. The more information we can extract from raw radar signals returned from targets, the more likely we will be to narrow down target identification. We must remember to emphasize that careful validation of processors and algorithms with targets identified by some other means is critical to their application in biological research. Unless we know exactly what processing algorithms are doing, we cannot accurately evaluate their performance. There is much to be done in this area.

The application of radar technology to conservation issues has just begun. Weather surveillance radar is already being used to examine the input and output of migrants at stopover areas, and this approach has great potential for identifying the geographical locations of important stopover areas for migrating birds. Future improvements such as dual-polarization for the WSR-88D will help researchers better discriminate insect echoes from those produced by birds in the atmosphere. Studies of migrant stopover ecology with high-resolution radar will aid in assessing suitability of potential stopover habitats because these radars have the ability to detect birds departing from different habitat types. New computer and software tools will make the task of data analysis much easier. Advancements in communications (fiber optic, wireless, and satellite) already allow a researcher to sit at a desk and download data from a remote radar site. Work is underway to link multiple marine radars and fuse the data that each is generating. These are exciting developments that suggest a very bright future for radar ornithology!

ACKNOWLEDGMENTS AND CREDITS

For the early mobile radar laboratory, the off-the-shelf radar was a Decca 150 (COTS) and the marine radar with the parabolic antenna was an LN-66 (Canadian Marconi Company). The high- resolution marine radar developed in 1998 (BIRDRAD) was based on a Furuno 2155 marine radar. For the thermal imaging project, the thermal imaging camera used was a

Radiance 1 (Raytheon-Amber, Calif.) and the marine radar used in the same study was a Pathfinder Model 3400 (Raytheon Inc., Manchester, N.H.). Our time-exposure device was a Video Peak Store (Colorado Video, Boulder).

INDEX

A

accounting, 78
adaptive radiation, 40
adjustment, 87
adults, 70, 71, 77
age, 3, 6, 17, 25, 28, 29, 30, 33, 45, 91
agents, 126, 131
aid, 165, 166
air, xi, 161
Air Force, 163, 165
allies, 24, 94
alpha, 124, 132
alternative, 29, 139, 141
alternative hypothesis, 139, 141
Amazon, 121, 124, 135
amendments, 18, 24
American Association for the Advancement of
    Science, 39
amorphous, 28
amphibia, 40
amphibians, 52, 56, 58, 60, 67, 78, 79, 86
amplitude, 55, 77, 87
analysts, 125
anatomy, 36, 41
angiosperms, 154
animals, vii, x, 1, 3, 17, 40, 89, 138, 153, 154, 157,
    158, 159
anomalous, 22, 23
anseriformes, viii, 1, 30, 41, 44
antenna, 163, 164, 165, 166
anthropology, 122
anus, 57
application, 166
aquatic systems, 50
arachnids, 52, 58, 78, 86
arthropods, 130
articulation, 29
asia, 11, 12, 23, 24, 33, 34, 38, 40

assessment, 88
assignment, 15, 17
assumptions, 19, 122
atlas, 41
atmosphere, 161, 166
attribution, 3, 18, 138
aura, 165
availability, 87

B

back, 21, 24, 26, 28, 31, 34
backlash, 122
banks, 53
beetles, 78
behavior, vii, 1, 10, 16, 23, 24, 26, 27, 41, 46, 52, 88,
    92, 124, 162
bell, 81, 88
benefits, viii, 2, 93
bias, 16, 17, 39
biodiversity, x, 89, 121, 122, 123, 124, 125, 126,
    127, 131, 132, 133, 134, 159
biomass, 53
biosphere, 160
bipedal, 23
body mass, 56
body size, 52, 89
breeding, ix, 91, 92, 93, 94, 164
burning, 125
butterfly, 142, 143

C

calyx, 154
candidates, 5, 31
case study, 19
category a, 55, 68, 69, 143
Central America, 160
changing environment, 92
circadian, 88, 89

circadian rhythm, 88
class intervals, 87
classification, 5, 38, 40, 45, 88
climate change, 138, 142, 143
climate warming, x, 137, 141
collisions, 163
colonization, 124, 138
combined effect, ix
communication, x, 18, 23, 121, 125, 160
community, viii, x, 49, 51, 85, 92, 93, 94, 123, 124,
153, 155, 156, 158, 159
compatibility, 122, 126
competition, viii, 49, 52, 87, 89, 91, 93, 94
complement, 4, 26
complex systems, 123
components, x, 51, 123, 153
composition, 53, 123
concentration, 50
conceptualization, 51
confidence, 5, 19, 21, 23, 33
configuration, 21, 163
Congress, 36, 42
consensus, x, 121, 123
conservation, ix, 121, 122, 123, 125, 126, 127, 131,
132, 134, 166
constraints, 24, 143
consumers, 50, 52
control, 3, 32, 163
convergence, viii, 2, 26, 31, 34, 42
coral, 134
coral reefs, 134
corolla, 154
correlation, 22, 33, 42, 81
correlation coefficient, 81
corridors, 138
crops, 126
crown, 28, 38
crustaceans, 52, 58, 81, 86

D

danger, 29, 32
data analysis, 166
data set, 33
database, 4, 125
dating, 28, 33
death, 3, 94
debates, 4
decay, 17, 22, 37
decisions, 21
defense, xi, 161
definition, 24
density, 162, 163, 164
deposition, 16

deposits, 8, 11, 16, 17, 23, 28, 33, 40, 45, 155
desert, 93, 142
desiccation, 87
destruction, 132
detection, 165, 166
deviation, 162
dichotomy, 32
diet, viii, x, 30, 47, 50, 52, 55, 66, 68, 70, 77, 78, 87,
88, 89, 93, 153, 154
diet composition, viii, 50
dietary, ix, 50, 52, 77, 87
differentiation, 90
digestion, 55
digestive process, 54
discipline, 2, 4, 5
discovery, 36, 43, 47
displacement, 55
distribution, vii, viii, ix, x, 1, 4, 16, 18, 23, 49, 88,
89, 92, 137, 138, 139, 140, 141, 142, 143, 162
diversity, viii, ix, x, 2, 15, 17, 22, 23, 26, 28, 30, 33,
34, 40, 50, 52, 55, 71, 72, 73, 74, 75, 92, 121,
122, 123, 124, 125, 126, 131, 132, 135, 143, 159
diving, 30, 50
dominance, viii, 2
Doppler, 164
download, 166
duplication, 34
duration, 8, 123

E

Earth Science, 35, 46, 47
East Asia, 12, 23, 33, 34
eating, ix, 47, 131, 132
ecological, vii, viii, x, 1, 25, 31, 47, 49, 51, 52, 53,
91, 94, 121, 124, 133, 134, 143
ecological systems, 134
ecologists, 123, 124, 134
ecology, vii, 1, 23, 34, 41, 51, 52, 53, 89, 123, 159,
166
ecosystem, 123, 124, 125
ecosystems, x, 25, 32, 52, 53, 123, 124, 135, 153,
154
education, 27
elephant, 3
elytra, 55
energy, 50, 143
enthusiasm, 122
environment, viii, x, 2, 50, 52, 54, 85, 86, 87, 88, 91,
121, 124, 125, 127, 131, 132, 134, 135
environmental factors, 32
equilibrium, 123, 124
erosion, 21
Europeans, 132

Index

171

Everglades, 91
evolution, vii, 1, 4, 16, 19, 20, 23, 25, 26, 28, 29, 32, 39, 42, 47, 52
evolutionary process, 143
exclusion, 52, 53, 124
exploitation, 87, 134
exposure, 19, 167
extinction, viii, 2, 32, 37, 124, 138, 142
eye, 130

F

facies, 16, 24, 43, 45
family, 10, 34, 37, 53, 69
fauna, 41, 50, 88
feeding, 26, 27, 30, 46, 50, 51, 52, 53, 56, 71, 81, 82, 86, 87, 88, 89, 91, 92, 94
feet, viii, 2, 6, 16, 22, 26, 29, 30, 31, 33, 42
femur, 44
fiber, 166
films, 164
firearm, 54
fish, 50, 67, 70, 77, 78, 79, 80, 81, 86, 87, 93
fishing, 52, 91
flame, 131
flexibility, 89
flight, xi, 39, 161, 162, 163, 164, 165, 166
floating, 50, 54, 76, 88
flood, 53, 54, 56, 92
fluvial, 16, 40, 43
focusing, x, 122, 153, 155
food, viii, 26, 50, 51, 52, 55, 56, 58, 59, 60, 61, 62, 63, 64, 65, 67, 68, 69, 70, 71, 86, 87, 88, 92, 93, 126
food intake, 92
forest ecosystem, 135
forest management, 122
forests, x, 54, 85, 122, 124, 126, 127, 131, 132, 154, 156
fossil, vii, viii, 1, 2, 3, 4, 6, 8, 10, 15, 16, 17, 22, 23, 24, 26, 28, 31, 32, 33, 34, 37, 39, 40, 41, 43, 44, 45
funding, 163

G

gait, 10, 26
geese, xi, 161
generation, 18, 164
geology, 22
gizzard, 47
glass, 55
global warming, ix, x, 137, 138, 142, 143
globalization, 123

goals, 41
grades, 28
grassland, 76, 88, 94, 126, 130
grouping, 3, 56
groups, viii, x, 2, 6, 12, 32, 34, 52, 61, 79, 86, 142, 153, 154
growth, 126, 127, 131, 132
guidance, 162
guidelines, 135
Guineans, 126

H

habitat, ix, 50, 51, 56, 57, 58, 59, 60, 61, 62, 63, 64, 65, 88, 91, 92, 94, 126, 131, 132, 138, 165, 166
health, 125
heart, 53
heterogeneity, 88
high risk, 142
high school, 161
holistic, 16
horses, 91
House, 148, 150
household, 126
human, x, 35, 121, 122, 124, 125, 131, 132
human activity, x, 121, 122, 125, 131
humanity, 124
hunter-gatherers, 124
hunting, 126
hurricanes, 161
hybrid, 159
hydrological, 87, 88
hypothesis, 23, 31, 139, 140, 141, 142, 143, 164

I

IBIS, 133
IBM, 92
identification, 22, 31, 51, 166
illumination, 165
imaging, 165, 166
imprinting, 21
incidence, 23
inclusion, 122
incompatibility, 52
Indian, 91
indication, 47
indicators, 50
indigenous, ix, 5, 121, 122, 125, 126, 127, 132, 135
indigenous knowledge, ix, 121, 122
infancy, 16
ingestion, 71
inheritance, 123

insects, ix, 50, 52, 58, 59, 60, 61, 63, 64, 68, 70, 71, 77, 78, 79, 80, 81, 86, 88, 130, 166
insight, 10, 33, 89, 132
institutions, 133
interaction, viii, xi, 49, 154, 156, 157, 158, 159
interval, 56, 71, 72, 73, 74, 75, 78, 79, 80, 81, 82
interviews, 126
intrinsic, 32, 123
inventories, 125
invertebrates, 6
IPCC, 138, 142
island, 53, 54, 122, 125, 126
isolation, ix, 5, 50, 52, 53, 87, 88, 89

J

jaw, 46
joining, 163
juveniles, 71, 77, 79, 80

K

killing, viii, 2

L

lagoon, 94
land, 54, 90, 122, 125, 126, 132, 162
land use, 132
landscapes, 122, 132
language, 134
Late Quaternary, 40
laws, 122
lens, 33, 165
lifestyle, x, 121, 126
lifetime, 17
light beam, 165
limitation, 156, 164
linear, ix, x, 82, 121
linear model, ix, 82
linkage, 87, 124, 131
location, 54
locomotion, 16, 30
low-level, xi, 161

M

mainstream, 2
maize, 135
mammal, 10, 42, 43, 44
management, 89, 92, 122, 125, 133, 135
management practices, 122
Mandarin, 139
marshes, 162
matrix, 85, 157

meanings, 51
measurement, 91, 92
measures, 124
media, 7, 146
Mesozoic, vii, viii, 1, 2, 6, 8, 9, 10, 11, 15, 16, 19, 21, 24, 25, 26, 28, 29, 30, 31, 32, 34, 35, 36, 37, 38, 39, 40, 41, 42, 43, 45, 46, 47
microclimate, 126
microhabitats, 52
migration, 93, 161, 162, 163, 164, 165
mines, 46
minority, 21
missions, 41
mitochondrial, 36
mitochondrial DNA, 36
mobility, 29
models, ix, 5, 81, 132
moisture, 161
mollusks, 52
money, 133
morning, 82
morphology, vii, viii, 1, 2, 5, 6, 11, 16, 18, 19, 20, 21, 22, 26, 29, 30, 31, 32, 34, 36, 37, 43, 52, 138
mosaic, x, 52, 88, 121, 124, 125, 126, 132
motion, 26
mountains, 132
movement, xi, 161

N

naming, 6, 7, 8, 10, 12, 19, 21
nation, 11, 122, 125
National Academy of Sciences, 47, 159
National Park Service, 92
natural environment, 87
nature conservation, 90
neck, 52
nesting, 52, 88
network, xi, 154, 157, 158, 161, 165
next generation, 164
nonequilibrium, 123
North America, 8, 9, 23, 24, 33, 35, 38, 43, 92, 94, 142, 143, 164
null hypothesis, 140

O

observations, ix, 15, 54, 88, 121, 156, 162
occupied territories, ix
offshore, 91
off-the-shelf, 166
omission, 19
one dimension, 51
operator, 163

opposition, 3
organism, 87, 138
orientation, 6
oversight, 21

P

Pacific, 94, 135
paleontology, 2, 5, 8, 22, 28
palmate, 14
parabolic, 163, 165, 166
paradigm shift, 124
parasites, x, 153, 154
partition, 51, 52
partnerships, 122, 127, 133
pastures, 54, 85
pedal, 18, 22, 25, 29, 30, 31, 32, 34
pelvic, 30
penguins, 30
percentage frequency, 55, 67, 68, 69, 70, 71
perception, 17, 122, 127
performance, 166
personal communication, 18, 23, 139
philosophy, ix, 121
phylogenetic, 9, 26, 29, 31, 37, 45
physiological, 50, 87
pica, 150
planning, ix, 121
plants, 61, 126, 138, 143, 158, 160
plastic, 54
plasticity, 47, 87
play, 50, 51, 87
polarization, 166
pollen, 155, 156
pollination, x, 153, 154, 155, 157, 158, 159, 160
pollinators, x, 126, 153, 154, 155, 156, 157, 158
poor, 21, 23
population, 123, 132
positive relationship, 164
post-extinction, 32
potato, 126, 132
power, 124
prediction, 26, 31
preference, ix, 50, 56, 85, 88
press, 3, 46, 154, 155, 158, 160
pressure, 54, 156
primates, 39
pristine, 124
probability, 55, 87, 89
probe, 30
productivity, 50, 53
program, 165
protocols, 127, 155
prototype, 164

public, 24
pulse, 92
pus, 20

R

radar, xi, 161, 162, 163, 164, 165, 166
radiation, 23, 33, 35, 37, 39, 40, 47
rail, 146
rain, 125, 159, 160, 161
rain forest, 159, 160
random, 15
range, x, 9, 19, 30, 32, 53, 56, 77, 78, 87, 89, 121, 141, 142, 143, 163, 164, 165
reality, 122
recognition, 3, 11, 29
reconcile, 122
recovery, x, 121, 124
reefs, 134
reflectivity, 162, 163, 164, 165, 166
regeneration, 131, 132
relationship, ix, 19, 24, 34, 76, 94, 121, 122, 123, 124, 125, 126, 162, 164
relatives, viii, 2
reproduction, x, 92, 153, 154
reproductive organs, 155
reptile, 44, 46
reputation, 3
research, 162, 163, 164, 166
researchers, 166
resistance, 158
resolution, 55, 165, 166
resources, viii, ix, x, 49, 50, 51, 52, 86, 87, 88, 89, 121, 123, 134, 138, 153, 154, 156, 159
rhythm, 56, 88
risk, 26, 138, 142, 143
river systems, 126
rivers, 53
robustness, 158
Royal Society, xi, 37, 39, 43, 46, 161
Russian, 15, 18, 46

S

sample, 9, 10, 55, 71, 72, 73, 75, 166
sand, 37
sandstones, 10
satellite, 166
scapula, 29
school, 124, 161
SCN, 138, 139, 140, 141, 144, 145, 146, 147, 148, 149, 150, 151
seabirds, 17
search, 26

Index

seasonality, 159
sedentary, ix, 124
sediment, 9, 16
seed, 47, 126
segregation, viii, 49, 51, 53, 88
selectivity, ix, 50, 56, 87
sensitivity, 163
series, 3, 13, 52, 89
services, 160
shape, 21
shorebirds, viii, 1, 2, 8, 16, 24, 26, 94, 162
short period, ix
signals, 166
similarity, 4, 18, 28, 55, 56, 87
sites, ix, 17, 23, 24, 52, 88
skeleton, 5, 17
skin, 6
Smithsonian, 36, 39, 44, 134
Smithsonian Institute, 134
soft drinks, 160
soft substrate, 13, 21
software, 56, 166
songbirds, 165
sorting, 46
South America, 14, 24, 35, 36, 41, 46, 53
Southern Hemisphere, 23
spatial, viii, 16, 50, 53, 88, 89, 162
spatial heterogeneity, 88
specialization, x, 40, 153, 154
species richness, 124
specific tax, 5
spectrum, ix, 50, 51, 52, 53, 56, 57, 58, 59, 60, 61,
   62, 63, 64, 65, 67, 164
speculation, 33
speed, 52, 165, 166
stages, 36, 132
stamens, 154
stereotypes, ix, 121, 124
stigma, 155
stomach, 54, 55, 56, 75, 89, 91
storms, 123
strategies, 50, 92, 134, 135
stress, 34, 123
subgroups, 51
subjective, 21
substrates, 21
supply, 88
surveillance, 162, 163, 164, 166
survival, 88, 142
survival rate, 88
sustainability, x, 121
sustainable development, 122
switching, 47

syndrome, 155, 158
synthesis, 8
systematics, 20, 36, 37
systems, 161

T

tar, 17
target identification, 166
targets, 162, 163, 165, 166
taxa, viii, 2, 5, 6, 9, 17, 22, 28, 29, 30, 31, 32, 33, 34,
   131
taxonomic, vii, viii, 1, 2, 17, 18, 23, 24, 25, 34, 40,
   44, 55, 56, 58, 59, 61, 62, 63, 64, 65, 86, 142
technological developments, 166
technology, 166
temperature, ix, x, 137, 138, 142
temporal distribution, 16
territory, 126
tetrapod, vii, 1, 32, 40
threat, xi, 161
thrush, 127, 149
tiger, 60, 68, 73, 79
time, 162, 163, 165, 167
time periods, 31
tissue, 17
tracking, 165
tradition, x, 15, 121, 122, 124, 125, 132
traditional practices, 122
traffic, 162, 163
trans, 12, 124, 162, 164, 165
transfer, 20, 50
transition, vii, 1, 42
transmission, 163, 165
transport, 17
transport processes, 17
trees, 126, 159
tropical forests, 123
tropical rain forests, 159
trust, 133

U

uncertainty, 15
undergraduate, 162
UNESCO, 93
universe, 123
USAID, 136

V

validation, 166
validity, 15, 21, 23
values, ix, 50, 55, 56, 63, 71, 72, 73, 74, 75, 76, 77,
   81, 85, 87, 88, 89, 164

variability, 53, 87
variable, 163
variation, 11, 52, 92, 166
vegetation, 53, 54, 68, 69, 70, 76, 85, 88, 125, 130
velocity, 164, 166
versatility, 89
vertebrates, 17, 35, 92, 130
visible, 162
vision, 163
voles, ix

W

walking, 22, 26, 43
water, 16, 17, 34, 50, 52, 53, 54, 55, 76, 85, 86, 87, 91, 94, 130

waterfowl, 93
weathering, 16, 21
web, 13, 14, 32
wild animals, 143
wilderness, 124, 125
wildlife, 50
wind, 166
windows, 17
wireless, 166
workers, 15, 18, 21

Y

yield, 10, 33